# Lecture Notes in Computer Science 11409

*Commenced Publication in 1973*
Founding and Former Series Editors:
Gerhard Goos, Juris Hartmanis, and Jan van Leeuwen

More information about this series at http://www.springer.com/series/7407

Yann Disser · Vassilios S. Verykios (Eds.)

# Algorithmic Aspects of Cloud Computing

4th International Symposium, ALGOCLOUD 2018
Helsinki, Finland, August 20–21, 2018
Revised Selected Papers

 Springer

*Editors*
Yann Disser (iD)
Technical University of Darmstadt
Darmstadt, Germany

Vassilios S. Verykios
Hellenic Open University
Patras, Greece

ISSN 0302-9743        ISSN 1611-3349   (electronic)
Lecture Notes in Computer Science
ISBN 978-3-030-19758-2      ISBN 978-3-030-19759-9   (eBook)
https://doi.org/10.1007/978-3-030-19759-9

LNCS Sublibrary: SL1 – Theoretical Computer Science and General Issues

This Springer imprint is published by the registered company Springer Nature Switzerland AG
The registered company address is: Gewerbestrasse 11, 6330 Cham, Switzerland

# Preface

The International Symposium on Algorithmic Aspects of Cloud Computing (ALGOCLOUD) is an annual event aiming to tackle the diverse new topics in the emerging area of algorithmic aspects of computing and data management in the cloud.

The aim of the symposium is to bring together international researchers, students, and practitioners to present research activities and results on topics related to algorithmic, design, and development aspects of modern cloud-based systems.

As in previous years, paper submissions were solicited through an open call for papers. ALGOCLOUD welcomes submissions on all theoretical, design, and implementation aspects of modern cloud-based systems. We are particularly interested in novel algorithms in the context of cloud computing, cloud architectures, as well as experimental work that evaluates contemporary cloud approaches and pertinent applications. We also welcome demonstration manuscripts, which discuss successful elastic system developments, as well as experience/use-case articles. Contributions may span a wide range of algorithms for modeling, practices for constructing, and techniques for evaluating operations and services in a variety of systems, including but not limited to, virtualized infrastructures, cloud platforms, datacenters, cloud-storage options, cloud data management, non-traditional key-value stores on the cloud, HPC architectures, etc.

Topics of interest addressed by this workshop include, but are not limited to:

- Distributed algorithms and mechanisms
- Algorithms, data structures, and computation
- Data science and machine learning
- Big data analytics and deep learning
- Networking, routing, and protocols
- Caching and load balancing
- Resource management and elasticity
- Search&retrieval and graph exploration
- Privacy and anonymization approaches
- Privacy-preserving record linkage
- Scale-up and -out for NoSQL and columnar databases
- Analysis of containerized applications
- Cloud deployment tools and their analysis
- Query languages and novel programming models
- Data structures and algorithms for eventually-consistent stores
- Scalable access structures and indexing for cloud data-stores
- NoSQL and schema-less data modeling and integration
- Consistency, replication, and partitioning CAP
- Transactional models and algorithms for cloud data-stores

ALGOCLOUD 2018 took place during August 20–21, 2018, in Helsinki, Finland. It collocated and was part of ALGO 2018 (August 20–24, 2018), the major annual congress that combines the premier algorithmic conference "European Symposium on Algorithms" (ESA) and a number of other specialized symposiums and workshops and a summer school, all related to algorithms and their applications, making ALGO the major European event for researchers, students, and practitioners in algorithms.

The Program Committee (PC) of ALGOCLOUD 2018 was delighted by the positive response to the call for papers. The diverse nature of papers submitted demonstrated the vitality of the algorithmic aspects of cloud computing. All submissions underwent the standard peer-review process and were reviewed by at least four PC members. The PC decided to accept 11 original research papers on a wide variety of topics that were presented at the workshop. We would like to thank the PC members for their significant contribution in reviewing process.

The program of ALGOCLOUD 2018 was complemented with a highly interesting keynote, entitled "Algorithms For and Against the Cloud," which was delivered by Roger Wattenhofer (ETH Zurich, Switzerland). We wish to express our sincere gratitude to this distinguished professor for the excellent keynote he provided.

We would like to thank all authors who submitted their research work in ALGOCLOUD. We also thank the Steering Committee for volunteering their time.

We hope that these proceedings will help researchers, students, and practitioners to understand and be aware of state-of-the-art algorithmic aspects of cloud computing, and that they will stimulate further research in the domain of algorithmic approaches in cloud computing in general.

August 2018

Yann Disser
Vassilios S. Verykios

# Organization

## Steering Committee

Spyros Sioutas      Ionian University, Greece
Peter Triantafillou      University of Glasgow, UK
Christos D. Zaroliagis      University of Patras, Greece

## Symposium Chairs

Yann Disser      TU Darmstadt, Germany
Vassilios S. Verykios      Hellenic Open University, Greece

## Program Committee

Alex Delis      University of Athens, Greece
Katerina Doka      National Technical University of Athens, Greece
Ahmed Eldawy      University of California Riverside, USA
Klaus-Tycho Foerster      University of Vienna, Austria
Aris Gkoulalas-Divanis      IBM Watson Health, USA
Dimitrios Karapiperis      Hellenic Open University, Greece
Eleftheria Katsiri      Democritus University of Thrace, Greece
Ulrich Meyer      Goethe University Frankfurt, Germany
Taneli Mielikainen      University of Helsinki, Finland
Paolo Missier      Newcastle University, UK
Nikolaos Nodarakis      University of Patras, Greece
Mourad Ouzzani      Qatar Computing Research Institute, HBKU, Qatar
Guido Proietti      University of L'Aquila, Italy
Juha Röning      University of Oulu, Finland
Yücel Saygin      Sabanci University, Turkey
Junho Shim      Sookmyung Women's University, South Korea
Elias C. Stavropoulos      Hellenic Open University, Greece
Przemysław Uznański      ETH Zurich, Switzerland
Dinusha Vatsalan      Data61 CSIRO, Australia

## Additional Reviewers

Panagiotis Kanellopoulos
Christina Karousatou
Panagiotis Liakos
Christos Makris
Evangelos Sakkopoulos

# Algorithms For and Against the Cloud (Keynote Talk)

Roger Wattenhofer

ETH Zurich, Switzerland
wattenhofer@ethz.ch
http://www.disco.ethz.ch

**Abstract.** Algorithms interact in two main ways with the cloud. There exist algorithms which are tailored *for the cloud*, for which the cloud is the perfect environment. Moreover, the cloud may also benefit from optimization algorithms, algorithms that make the *cloud more efficient*. The AlgoCloud program features papers which roughly fit one of the two, and there will be a few examples in the first part of the presentation.

The first example [2] studies how to solve a large computational problem, represented by a graph, by partitioning it into two or more smaller parts. Each part is solved on a single processor, in parallel. The vertices of a component are simulated on a single processor whereas edges between two vertices in different components are handled by the two processors responsible for the two components by exchanging messages. A natural objective of designing such a partition is to reduce the inter-processor communication as it is the expensive part in terms of runtime. We argue that an input graph should be partitioned by means of a balanced vertex separator (and not a balanced edge cut), since vertex separators are often more efficient. We sketch how to find a small balanced vertex separator (if one exists) in almost linear time.

Next we show some examples well suited for the cloud. In particular we discuss GPS coarse time navigation [1] and GPS spoofing, as well as online matching [3]. The online matching problem is an example of an online problem which only becomes feasible if we allow for delaying decisions. So the cost of the algorithm two-fold: (i) the cost of the quality of the matching, and (ii) the waiting time until something is matched. To the best of our knowledge, [3] was the first online all pairs matching.

So far for algorithms *for the cloud*. The second part of the presentation is about algorithms *against* the cloud. Recently, blockchains [4] are hyped to be a cloud competitor, sometimes even a cloud killer. First we quickly discuss some of the basics of blockchains, with eMoney and eVoting as examples. Then we want to know whether there is some truth to whether blockchains are going to threaten the successful cloud paradigm. We discuss several cloud vs. blockchain angles: Should one trust a large (usually trusted) corporation, or rather thousands (of usually untrusted) individuals? What about energy consumption? And what should we think if a corporation offers to run a blockchain in their cloud?

# References

1. Bissig, P., Eichelberger, M., Wattenhofer, R.: Fast and robust GPS fix using one millisecond of data. In: 16th ACM/IEEE International Conference on Information Processing in Sensor Networks (IPSN), Pittsburgh, Pennsylvania, USA, April 2017
2. Brandt, S., Wattenhofer, R.: Approximating small balanced vertex separators in almost linear time. In: Ellen, F., Kolokolova, A., Sack, J.R. (eds.) WADS 2017. LNCS, vol. 10389, pp. 229–240. Springer, Cham (2017). https://doi.org/10.1007/978-3-319-62127-2_20
3. Emek, Y., Kutten, S., Wattenhofer, R.: Online matching: haste makes waste! In: 48th Annual Symposium on the Theory of Computing (STOC), Cambridge, Massachusetts, USA, June 2016
4. Wattenhofer, R.: Blockchain Science: Distributed Ledger Technology. Inverted Forest Publishing (2016–2019)

# Contents

# Minimization of Testing Costs in Capacity-Constrained Database Migration

K. Subramani[1]([⊠]), Bugra Caskurlu[2], and Alvaro Velasquez[3]

[1] LCSEE, West Virginia University, Morgantown, WV, USA
k.subramani@mail.wvu.edu
[2] CE, TOBB University of Economics and Technology, Ankara, Turkey
caskurlu@gmail.com
[3] RISC, Air Force Research Laboratory, Rome, NY, USA
alvaro.velasquez@us.af.mil

**Abstract.** Database migration is an ubiquitous need faced by enterprises that generate and use vast amount of data. This is due to database software updates, or from changes to hardware, project standards, and other business factors [1]. Migrating a large collection of databases is a way more challenging task than migrating a single database, due to the presence of additional constraints. These constraints include capacities of shifts, sizes of databases, and timing relationships. In this paper, we present a comprehensive framework that can be used to model database migration problems of different enterprises with customized constraints, by appropriately instantiating the parameters of the framework. We establish the computational complexities of a number of instantiations of this framework. We present fixed-parameter intractability results for various relevant parameters of the database migration problem. Finally, we discuss a randomized approximation algorithm for an interesting instantiation.

## 1 Introduction

The database migration problem entails the movement of data between different databases. Such migration is often necessary due to database software updates, or from changes to hardware or project standards, and other business factors [1]. As per established rules of software reliability, when a database is migrated, every application that is dependent upon it *must* be tested (i.e., run through regression suites [2]). It is known that testing an application is an expensive aspect of maintaining the application [3]. Consequently, the principal goal in the migration process is to minimize the application testing cost [4]. This is a

K. Subramani—This research was supported in part by the Air Force Research Laboratory Information Directorate, through the Air Force Office of Scientific Research Summer Faculty Fellowship Program and the Information Institute®, contract numbers FA8750-16-3-6003 and FA9550-15-F-0001.

© Springer Nature Switzerland AG 2019
Y. Disser and V. S. Verykios (Eds.): ALGOCLOUD 2018, LNCS 11409, pp. 1–12, 2019.
https://doi.org/10.1007/978-3-030-19759-9_1

pervasive issue in cloud computing clusters, where a pay-per-use infrastructure and unpredictable workloads necessitate frequent allocation and movement of data [5]. Thus, there is a pressing need for efficient procedures to minimize the resource overhead involved in data migration. Interest in this area has also been fueled in recent years by the massive generation of data in what is now being called the Big Data age [6]. Consequently, there has been a proliferation of resource-intensive data centers and the adoption of cloud computing and storage as a service in order to manage said data [7].

While the ubiquity of computational and storage capabilities of cloud computing are undeniable, there remain open challenges with regards to resource allocation and data management [8]. In fact, the migration of data in data centers remains a significant problem, due to the massive throughput of data and the limited bandwidth of communication channels [9]. This problem is further exacerbated by the overhead incurred from retesting applications after data migration and by Quality-of-Service (QoS) requirements, which demand minimal interruptions to end-user applications [10]. As such, minimizing the cost associated with bandwidth-constrained database migration is of great interest. Patil et al. [11] introduced the first systematic study on the database migration problem, and proved that the problem is **NP-hard** for some special cases. They provided an integer programming formulation, which can only be used for small instances, and a greedy heuristic that can be used for the large instances of the problem. In this paper, we define a very general framework that subsumes the model in [11]. Our framework accommodates the modelling of database migration needs of various enterprises with customized constraints. We present hardness results for all the models in our framework, as well as fixed-parameter intractability for various relevant parameters. We also present a randomized approximation algorithm for a simple but interesting special case of the problem.

The rest of the paper is organized as follows:

- In Sect. 2, we formally define the capacity-constrained database migration (CCDM) problem and introduce the notation used in the paper.
- In Sect. 3, we study the computational complexity of the (CCDM) problem and prove that the problem is **NP-hard** for all the models of the problem defined in Sect. 2.
- In Sect. 4, we study fixed parameter tractability of the CCDM problem with respect to the four most relevant parameters of the problem.
- In Sect. 5, we present a randomized $(\frac{3}{2} + \epsilon)$-approximation algorithm for a special case of the CCDM problem.
- In Sect. 6, we summarize our contributions and point out avenues for future research.

## 2   Notations and Problem Formulation

In this section, we provide a formal definition of the framework for the capacity constrained database migration problem (CCDM) and introduce the terminology used in this paper.

Assume we have a collection of $m$ applications $\mathcal{A} = \{A_1, A_2, \ldots, A_m\}$ and $n$ databases $\mathcal{B} = \{B_1, B_2, \ldots, B_n\}$, with each application calling one or more databases. The call relationship is stored in the $n \times m$ matrix $\mathbf{D} = [d_{ij}]$, where

$$d_{ij} = \begin{cases} 1, & \text{if application } A_i \text{ calls database } B_j \\ 0, & \text{otherwise.} \end{cases}$$

The matrix $\mathbf{D} = [d_{ij}]$, which represents a bipartite graph as shown in Fig. 1, is part of the input. Associated with the set of applications $\mathcal{A}$, is a cost-vector $\mathbf{c} = [c_1, c_2, \ldots, c_n]^T$, where $c_i$ represents the cost of testing application $A_i$ once. For each application $A_i$, we let $x_i$ be an integer variable that denotes the number of times $A_i$ will have to be tested in the migration schedule. The size-vector $\mathbf{w} = [w_1, w_2, \ldots, w_m]^T$ represents the size of databases, with $w_i$ representing the size of database $B_i$.

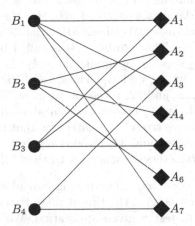

**Fig. 1.** The bipartite graph shows the relationship between the databases and the applications. The nodes in the left partition, represent the databases in the system, while the nodes in the right partition, represent the applications. An edge $(b, a)$ exists in the graph if application $a$ calls database $b$. This means application $a$ must be tested immediately after database $b$ migrates. We note that each database is associated with a nonempty set of applications, and each application is associated with a nonempty set of databases.

In the CCDM problem, the set of databases $\mathcal{B}$ is to be clustered into disjoint subsets which we call shifts. The databases in each shift are migrated at the same time. When a shift of databases migrates, each application that calls at least one database in that shift needs to be tested immediately. For example, if the set of databases called by an application $A_i$ are scheduled to 5 distinct shifts, then application $A_i$ is to be tested 5 times throughout the migration process, i.e., $x_i = 5$. The cumulative size of the databases migrated in shift $i$ (i.e., the size of shift $i$) is denoted by $l_i$. The shift size-vector $\mathbf{l} = [l_1, l_2, \ldots, l_m]^T$ is also part of the input. In the worst case, when the size of each shift is smaller than the sum of the sizes of any two databases, we may have to assign each database to a separate shift.

Thus, the input to the CCDM problem must contain the 4-tuple $\langle \mathbf{c}, \mathbf{w}, \mathbf{D}, \mathbf{l} \rangle$. For instance, consider the following 4-tuple:

$$\left\langle \begin{pmatrix} 1 \\ 1 \\ 1 \\ 2 \\ 2 \\ 3 \\ 3 \end{pmatrix}, \begin{pmatrix} 5 \\ 7 \\ 10 \\ 12 \end{pmatrix}, \begin{pmatrix} 1\,0\,1\,0\,1\,0\,1 \\ 0\,1\,0\,1\,0\,1\,0 \\ 1\,1\,0\,0\,1\,0\,0 \\ 0\,0\,1\,0\,0\,0\,1 \end{pmatrix}, \begin{pmatrix} 10 \\ 15 \\ 12 \\ 7 \\ 28 \\ 34 \end{pmatrix} \right\rangle$$

In this example, we have seven applications and four databases, since the matrix $\mathbf{D}$ has seven columns and four rows. We label the applications as $A_1, A_2, \ldots, A_7$, and the databases as $B_1, B_2, B_3, B_4$. Matrix $\mathbf{D}$ indicates which applications call which databases. We observe that database $B_1$ is called by applications $A_1, A_3, A_5$, and $A_7$. The database $B_2$ is called by applications $A_2, A_4$, and $A_6$. The database $B_3$ is called by applications $A_1, A_2$, and $A_5$, and database $B_4$ is called by applications $A_3$, and $A_7$. The sizes of the databases $B_1, B_2, B_3$, and $B_4$ are given as $5, 7, 10$, and $12$, respectively, in $\mathbf{w}$. The database migration operation in this example is to be completed in at most 6 shifts since $|\mathbf{l}| = 6$ in the input. The cumulative size of the databases that migrate in each shift is constrained by $10, 15, 12, 7, 28$, and $34$, respectively, in $\mathbf{l}$.

There are several parameters associated with the CCDM problem:

(i) Application Testing cost $(\alpha)$ - The testing cost of an application depends primarily on two factors, viz., the time it takes to test the application, and the skills required to test a given application. We consider the following three cost models associated with testing an application in order to take these factors into account:

   (a) Constant (*const*) - In this model, the cost of testing each application is the same and is equal to some known fixed constant $C$. Note that this model captures the scenario in which each application requires the same level of skills and takes roughly the same time to test.

   (b) Proportional (*prop*) - In this model, the cost of testing an application after migrating a corresponding database is proportional to the sizes of the migrated databases it calls. Note that this model captures the situation where each application in the system requires the same level of skills, but the testing time may vary from application to application.

   (c) Arbitrary (*arb*) - In this model, there is no relation among the costs of testing different applications. This model captures the problem of companies that use application software from several different companies such that each application requires personnel with different skill sets.

(ii) Size of Databases $(\beta)$ - The size of a database is a factor in the amount of time required for its migration. This is because the database migration operation must read the database at the original location, write it at the new location, and then delete the original database. Consider a bank that maintains data for credit card accounts, savings accounts, and checking accounts in different databases. Typically, a bank will have more customers with a checking account than a savings account. Similarly, the number of credit card customers will be significantly more than the number of savings account customers, since a typical customer has one savings account but several credit cards. This means we need two size models associated with the databases:

(a) Constant (*const*) - In this model, all databases have the same size and are equal to some fixed constant $W$. This allows us to model the database migration operation for companies whose databases have more or less the same size.

(b) Arbitrary (*arb*) - In this model, the sizes of the databases are arbitrary. This lets us model the database migration operation for companies whose databases may significantly vary in size.

(iii) Shift size ($\gamma$) - During the database migration operation, some parts of the database will be inaccessible. For some companies, there is no ideal time to make a database unavailable. For instance, Facebook and Youtube have users all over the world which means that the database access rate is roughly uniform. In this case, regardless of when a database becomes inaccessible, there will be a subset of users who cannot access the database until the migration is complete. It is critical for these companies to perform the database migration operation in small shifts to minimize user dissatisfaction. For companies (e.g., banks) that operate during regular business hours, it is preferable for the databases to be unavailable when the companies are closed rather than when they are open. In order to model the needs of several different companies, we use two size models associated with shifts.

(a) Uniform (*unif*) - In this model, the total size of the databases migrated in the same shift is the same and is equal to a constant $L$, for all shifts (i.e., $\mathbf{l} = \langle L, L, \ldots, L \rangle$). We note that this model is better suited for the database migration needs of companies that have uniform database access rates.

(b) Non-uniform (*non-unif*) - In this model, the total size of the databases migrated in each shift is arbitrary. We note that this model is better suited for the database migration needs of companies that have non-uniform database access rates.

Thus, a model of the capacity-constrained database migration problem has three parameters, and it is specified as a triple $\langle \alpha \mid \beta \mid \gamma \rangle$. For instance, $\langle arb \mid const \mid unif \rangle$ refers to the capacity-constrained database migration problem in which the application testing costs are arbitrary, all databases have the same size, and the shift sizes are uniform. For notational convenience we use $*$ as an entry of the triple when we present a statement that is true for all the models for that entry. For instance, the notation $\langle arb \mid * \mid * \rangle$ refers to all 4 models of the CCDM problem in which the application testing costs are arbitrary. The following is formal definition of the CCDM problem:

**CCDM:** *Given a 4-tuple* $\langle \mathbf{c}, \mathbf{w}, \mathbf{D}, \mathbf{l} \rangle$, *cluster the databases into shifts so that the total application testing cost is minimized, while respecting the shift size constraints.*

Given that we have 3 different models for application testing costs, 2 different models for sizes of databases, 2 different models for shift sizes; the CCDM problem formulation gives us a framework with a total of 12 different models each of which is suitable for the database migration needs of different companies.

## 3  Computational Complexity of the CCDM Problem

The formulation given for the database migration problem in [11] corresponds to our CCDM problem under the model $\langle arb \mid arb \mid unif \rangle$. In [11], it is proven that the CCDM problem is **NP-hard** under the model $\langle arb \mid * \mid * \rangle$. In this section, we strengthen the result in [11] via Theorem 1, which states that the CCDM problem is **NP-hard** for all the models in our framework, even under the restriction that there are only 2 shifts and each application calls at most 2 databases.

**Theorem 1.** *The CCDM problem in models* $\langle * \mid * \mid * \rangle$ *is* **NP-hard** *even under the restrictions that there are only two shifts, and each application calls at most two databases.*

*Proof.* Since the set of instances of the capacity-constrained database migration problem under the model $\langle const \mid const \mid unif \rangle$ is a subset of the instances of any model captured by the notation $\langle * \mid * \mid * \rangle$, all we need to do is to prove that the CCDM problem is **NP-hard** under the model $\langle const \mid const \mid unif \rangle$, when there are only two shifts, and each application calls at most two databases. We will do this via a polynomial time reduction from the classical MINIMUM-BISECTION problem [12], whose definition is given below.

**Definition 1 (MINIMUM-BISECTION).** *Given an undirected graph $G = (V, E)$, partition $V$ into two subsets $V_1$ and $V_2$ of equal size such that the number of edges with one endpoint in $V_1$ and one endpoint in $V_2$ are minimized. It is assumed $|V|$ is even.*

For a given instance $G = (V, E)$ of the MINIMUM-BISECTION problem, we construct the corresponding CCDM instance in model $\langle const \mid const \mid unif \rangle$ as follows:

- For every vertex $i$ of the graph of the MINIMUM-BISECTION instance, the CCDM instance has a corresponding database $B_i$ with unit size, i.e., $w_i = 1$,
- For every edge $e = (i, j)$ of the graph of the MINIMUM-BISECTION instance, the CCDM instance has a corresponding application $A_e$ with unit application testing cost($c_e = 1$) that calls databases $B_i$ and $B_j$,

– The CCDM instance has only two shifts and the size of each shift is $\frac{|V|}{2}$.

Notice that the constructed CCDM instance has $|V|$ databases and $|E|$ applications such that each application calls exactly two databases. In any feasible solution to the CCDM instance exactly $k = \frac{|V|}{2}$ databases are assigned to the first shift and the remaining $k$ databases are assigned to the second shift. If both of the databases called by an application $A_i$ are assigned to the same shift then application $A_i$ needs to be tested only once ($x_i = 1$) in the database migration process, and it needs to be tested twice ($x_i = 2$) otherwise. Let $d$ denote the number of applications that calls one database from each shift, and thus needs to be tested twice. Then, the total application testing cost of the CCDM instance is $|V| + d$, since $d$ of the $|E|$ applications are to be tested twice whereas all other applications are to be tested only once. Since $|V|$ is fixed, the optimal solution to the CCDM instance is the one that minimizes $d$.

Given a solution to constructed CCDM instance, consider the following solution to the given MINIMUM-BISECTION instance:

– If database $B_i$ of the constructed CCDM instance is assigned to the first shift, then assign vertex $i$ of the MINIMUM-BISECTION instance to $V_1$,
– If database $B_i$ of the constructed CCDM instance is assigned to the second shift, then assign vertex $i$ of the MINIMUM-BISECTION instance to $V_2$.

Notice that an edge $e = (i, j)$ of the given MINIMUM-BISECTION instance has one endpoint in $V_1$ and one endpoint in $V_2$ if and only if the databases $B_i$ and $B_j$ of the CCDM instance are assigned to different shifts and thus the application $A_e$ is to be tested twice. Therefore, the number of edges with one endpoint in $V_1$ and one endpoint in $V_2$ of the given MINIMUM-BISECTION instance is equal to the number of applications that needs to be tested twice in the constructed CCDM instance, which is denoted by $d$.                    □

## 4    Fixed Parameter Intractability of the CCDM Problem

In this section, we study the fixed parameter tractability of the CCDM problem for various parameters of the problem such as the number of applications, the number of shifts, the maximum number of databases that is called by an application, and the maximum number of applications that calls a database.

Notice that Theorem 1 establishes fixed parameter intractability for all 12 models of the CCDM problem, where the parameter is the number of shifts, or the maximum number of databases that is called by an application. In the rest of the section, we prove intractability results for the remaining parameters.

Theorem 2 establishes fixed parameter intractability for all 6 models of the CCDM problem captured by the notation $\langle * \mid arb \mid * \rangle$, when the parameter is the number of applications. Theorem 3 establishes fixed parameter intractability for all 6 models of the CCDM problem captured by the notation $\langle * \mid * \mid non - unif \rangle$, when the parameter is the maximum number of applications that calls a database.

**Theorem 2.** *The CCDM problem is **NP-hard** under the 6 models captured by the notation $\langle * \mid arb \mid * \rangle$, even when the number of applications $|\mathcal{A}|$ is 2.*

*Proof.* Since the set of instances of the database migration problem under the model $\langle const \mid arb \mid unif \rangle$ is a subset of the instances of any model captured by the notation $\langle * \mid arb \mid * \rangle$, all we need to do is to prove that the CCDM problem is **NP-hard** under the model $\langle const \mid arb \mid unif \rangle$, when $|\mathcal{A}|$ is 2. We will do that via a polynomial reduction from the classical PARTITION problem [12], whose definition is given below.

**Definition 2 (PARTITION).** *Given a multiset $S$ of positive integers, can $S$ be partitioned into two subsets $S_1$ and $S_2$ such that the sum of the numbers in $S_1$ equals the sum of the numbers in $S_2$?*

Given a PARTITION instance $S = \{s_1, s_2, \ldots, s_n\}$, we construct a corresponding CCDM instance under the model $\langle const \mid arb \mid unif \rangle$ with 2 applications as follows:

- For every integer $s_i$ in the multiset $S$ of the PARTITION instance, the CCDM instance has a corresponding database $B_i$ with size $s_i$, i.e., $w_i = s_i$,
- The CCDM instance has two applications $A_1$ and $A_2$ with unit testing costs, each of which calls all of the $n$ databases,
- The CCDM instance has sufficiently many shifts and the size of each shift is $\frac{\sum_{i=1}^{n} s_i}{2}$.

Since both of the applications of the CCDM instance calls all the databases, the total application testing cost is two times the number of shifts with at least one database assigned. Thus, the CCDM instance has a solution with total application testing cost of 4 if the databases can be clustered into two shifts. The theorem holds since the databases can be clustered into two shifts if and only if the answer to the PARTITION instance is yes.                                               □

**Theorem 3.** *CCDM is strongly **NP-hard** for the 6 models captured by the notation $\langle * \mid * \mid non - unif \rangle$, even if each database is called by two applications.*

*Proof.* Since the set of instances of the capacity-constrained database migration problem under the model $\langle const \mid const \mid non - unif \rangle$ is a subset of the instances of any model captured by the notation $\langle * \mid * \mid non - unif \rangle$, all we need to do is to prove that the CCDM problem is **NP-hard** under the model $\langle const \mid const \mid non - unif \rangle$, when each database is called by at most two applications. We will do that via a polynomial reduction from the classical CLIQUE problem, whose definition is given below.

**Definition 3 (CLIQUE).** *Given a graph $G = (V, E)$ and an integer $k$, is there a fully connected subgraph $G' \subseteq G$ consisting of $k$ vertices?*

Given a CLIQUE instance $\langle G = (V, E), k \rangle$, we construct a corresponding CCDM instance under the model $\langle const \mid const \mid non - unif \rangle$ as follows:

- For every vertex $v_i$ of the CLIQUE instance, the CCDM instance has a corresponding application $A_i$ with unit application testing cost, i.e., $c_i = 1$.
- For every edge $e$ of the CLIQUE instance, the CCDM instance has a corresponding database $B_e$ with unit size, i.e., $w_i = 1$,
- For every edge $e = (v_i, v_j)$ of the CLIQUE instance, the applications $A_i$ and $A_j$ of the CCDM instance calls database $B_e$. Notice that each database is called by exactly 2 applications.
- The CCDM instance has $|E| - k(k-1)/2 + 1$ shifts. The size of the first shift is $k(k-1)/2$, and the size of each of the remaining $|E| - k(k-1)/2$ shifts is 1.

We next show that there is a $k$-clique in $G$ if and only if our CCDM instance yields a solution with a total application testing cost of $2|E| - k^2 + 2k$.

Let $c^*$ denote the cost of the minimum-cost solution for our CCDM instance. Since each database is connected to 2 applications, we have $c^* > 2(|E| - k(k-1)/2) = 2|E| - k^2 + k$. This follows from the fact that each database placed in one of the $|E| - k(k-1)/2$ shifts with capacity of 1 must have both of its connected applications tested, leading to a test cost of $2(|E| - k(k-1)/2) = 2|E| - k^2 + k$. For the shift with capacity $k(k-1)/2$, we have a total application test cost of $k$ if and only if the vertices in $G$ corresponding to the databases in this shift induce a clique of size $k$. This follows trivially from the fact that the number of edges in a $k$-clique is $k(k-1)/2$. Thus, $c^* \geq 2|E| - k^2 + 2k$ and $c^* = 2|E| - k^2 + 2k$ if and only if $G$ has a clique of size $k$. Since all values in the reduction are polynomially bounded, it follows that this problem is **NP-hard** in the strong sense.     □

## 5   Approximation Algorithm for a Special Case of the CCDM Problem

In this section, we present Algorithm 5.1 for the CCDM problem under the model $\langle const \mid const \mid unif \rangle$, when there are only two shifts and each application calls at most two databases. Algorithm 5.1 is a randomized $(\frac{3}{2} + \epsilon)$−approximation algorithm for any given $\epsilon > 0$ by Theorem 4.

---

**Function** MIN-TEST-COST($\langle \mathbf{c}, \mathbf{w}, \mathbf{D}, \mathbf{l} \rangle$, $\epsilon$)

1: **if** $n < 1 + \frac{1}{2\epsilon}$ **then**
2:     Find optimal solution by brute force
3: **else**
4:     Select half of the databases by simple random sampling without replacement
5:     Assign the selected databases to the first shift
6:     Assign the remaining databases to the second shift
7: **end if**

---

**Algorithm 5.1.** Randomized $(\frac{3}{2} + \epsilon)$−approximation algorithm for CCDM problem under the model $\langle const \mid const \mid unif \rangle$, when there are only two shifts and each application calls at most two databases.

**Theorem 4.** *For any given $\epsilon > 0$, Algorithm 5.1 returns a solution whose total application testing cost is at most $(\frac{3}{2} + \epsilon)$ times that of the optimum, for the CCDM problem under the model $\langle const \mid const \mid unif \rangle$, when there are only two shifts and each application calls at most two databases.*

*Proof.* Since Algorithm 5.1 finds the optimal solution in polynomial time by brute force if $n < 1 + \frac{1}{2\epsilon}$, in the rest of the proof, we will assume the contrary, i.e., $\epsilon \geq \frac{1}{2n-2}$.

Since there are only two shifts, each application $A_i$ is to be tested only once or twice. If both of the databases that are called by $A_i$ are assigned to the same shift or if it calls only one database it will be tested once, otherwise it will be tested twice. Let $C$ denote the application testing cost of any application. Then the total application testing cost is $C$ times the total number of application tests. Note that $m \cdot C$ is a lower bound for the cost of the optimal solution to this CCDM instance, since each application is to be tested at least once. (Recall that $m$ is the number of applications and $n$ is the number of databases.)

Let $X_i$ be a random variable denoting the number of times application $A_i$ is to be tested with respect to the migration schedule generated by Algorithm 5.1. The total cost of the migration schedule generated by Algorithm 5.1 is then $C \cdot \sum_{i=1}^{m} X_i$. To complete the proof, all that we need to do is to show that $\mathbb{E}(\sum_{i=1}^{m} X_i) \leq (\frac{3}{2} + \epsilon) \cdot m$. Since $\mathbb{E}(\sum_{i=1}^{m} X_i) = \sum_{i=1}^{m} \mathbb{E}(X_i)$ due to linearity of expectations, it suffices to show that $\mathbb{E}(X_i) \leq (\frac{3}{2} + \epsilon)$ for any $i$. If $A_i$ calls only one database then this inequality is trivially satisfied since $\mathbb{E}(X_i) = 1$. So, we focus on applications that calls exactly two databases.

Let $A_i$ be an application that calls databases $B_j$ and $B_k$. Let $E_j$ and $E_k$ denote the events that databases $B_j$ and $B_k$ are assigned to the first shift respectively. Accordingly, $\overline{E_j}$ and $\overline{E_k}$ are the events that the databases $B_i$ and $B_j$ are assigned to shift 2.

It follows that

$$
\begin{aligned}
\mathbb{E}(X_i) &= 1 \cdot \mathbf{Pr}((E_j \cap E_k) \cup (\overline{E_j} \cap \overline{E_k})) + 2 \cdot \mathbf{Pr}((E_j \cap \overline{E_k}) \cup (\overline{E_j} \cap E_k)) \\
&= \mathbf{Pr}(E_j \cap E_k) + \mathbf{Pr}(\overline{E_j} \cap \overline{E_k}) + 2 \cdot \mathbf{Pr}(E_j \cap \overline{E_k}) + 2 \cdot \mathbf{Pr}(\overline{E_j} \cap E_k) \\
&= \mathbf{Pr}(E_j)\mathbf{Pr}(E_k|E_j) + \mathbf{Pr}(\overline{E_j})\mathbf{Pr}(\overline{E_k}|\overline{E_j}) \\
&= 2\left(\mathbf{Pr}(E_j)\mathbf{Pr}(\overline{E_k}|E_j) + \mathbf{Pr}(\overline{E_j})\mathbf{Pr}(E_k|\overline{E_j})\right) \\
&= \frac{1}{2} \cdot \frac{\frac{n}{2}-1}{n-1} + \frac{1}{2} \cdot \frac{\frac{n}{2}-1}{n-1} + 2\left(\frac{1}{2} \cdot \frac{\frac{n}{2}}{n-1} + \frac{1}{2} \cdot \frac{\frac{n}{2}}{n-1}\right) \\
&= \frac{3}{2} + \frac{1}{2n-2} \\
&\leq \frac{3}{2} + \epsilon, \text{ as desired.}
\end{aligned}
$$

# 6  Conclusion and Future Research Directions

This paper presented a general framework that is suitable for modelling the database migration requirements of a variety of enterprises. We showed that the CCDM problem is **NP-hard** for all the models considered, even under the very restricted scenario, where there are only 2 shifts and each application calls at most 2 databases. We also studied the parameterized complexity of the CCDM problem for four relevant parameters and presented fixed parameter intractability results for all of them. Finally, we presented a $(\frac{3}{2} + \epsilon)$-approximation algorithm for an interesting but a quite restricted special case of the CCDM problem. Every model of the CCDM problem is an interesting combinatorial optimization problem by itself, and it would be interesting to know for which models of the CCDM problem there are low factor approximation algorithms, and for which models there are not. From our perspective, the following avenues of research are interesting:

1. Derandomizing the randomized approximation algorithm.
2. Designing approximation algorithms and/or obtaining inapproximability results for all the models of the CCDM problem.

# References

1. Ravikumar, Y.V., Krishnakumar, K.M., Basha, N.: Oracle database migration. Oracle Database Upgrade and Migration Methods, pp. 213–277. Apress, Berkeley (2017). https://doi.org/10.1007/978-1-4842-2328-4_5
2. Harrold, M.J., et al.: Regression test selection for java software. In: ACM SIGPLAN Notices, vol. 36, pp. 312–326. ACM (2001)
3. Vergilio, S.R., Maldonado, J.C., Jino, M., Soares, I.W.: Constraint based structural testing criteria. J. Syst. Softw. **79**(6), 756–771 (2006)
4. Eric Wong, W., Horgan, J.R., Mathur, A.P., Pasquini, A.: Test set size minimization and fault detection effectiveness: a case study in a space application. J. Syst. Softw. **48**(2), 79–89 (1999)
5. Elmore, A.J., Das, S., Agrawal, D., El Abbadi, A.: Zephyr: live migration in shared nothing databases for elastic cloud platforms. In: Proceedings of the 2011 ACM SIGMOD International Conference on Management of data, pp. 301–312. ACM (2011)
6. Lohr, Steve: The age of big data. New York Times **11**, 2012 (2012)
7. Bahrami, M., Singhal, M.: The role of cloud computing architecture in big data. In: Pedrycz, W., Chen, S.-M. (eds.) Information Granularity, Big Data, and Computational Intelligence. SBD, vol. 8, pp. 275–295. Springer, Cham (2015). https://doi.org/10.1007/978-3-319-08254-7_13
8. Nascimento, D.C., Pires, C.E., Mestre, D.: Data quality monitoring of cloud databases based on data quality SLAs. In: Trovati, M., Hill, R., Anjum, A., Zhu, S.Y., Liu, L. (eds.) Big-Data Analytics and Cloud Computing, pp. 3–20. Springer, Cham (2015). https://doi.org/10.1007/978-3-319-25313-8_1
9. Ping, L., Zhang, L., Liu, X., Yao, J., Zhu, Z.: Highly efficient data migration and backup for big data applications in elastic optical inter-data-center networks. IEEE Network **29**(5), 36–42 (2015)

10. Xiaonian, W., Deng, M., Zhang, R., Zeng, B., Zhou, S.: A task scheduling algorithm based on qos-driven in cloud computing. Procedia Comput. Sci. **17**, 1162–1169 (2013)
11. Patil, S., et al.: Minimizing testing overheads in database migration lifecycle. In: COMAD, p. 191 (2010)
12. Papadimitriou, C.H.: Computational Complexity. Addison-Wesley, New York (1994)

# Community Detection via Neighborhood Overlap and Spanning Tree Computations

Ketki Kulkarni[1], Aris Pagourtzis[2]([✉]), Katerina Potika[1], Petros Potikas[2], and Dora Souliou[2]

[1] San Jose State University,
One Washington Square, San Jose, CA 95192, USA
{ketki.kulkarni,katerina.potika}@sjsu.edu
[2] School of Electrical and Computer Engineering,
National Technical University of Athens, 15780 Zografou, Greece
{pagour,ppotik}@cs.ntua.gr, dsouliou@gmail.com

**Abstract.** Most social networks of today are populated with several millions of active users, while the most popular of them accommodate way more than one billion. Analyzing such huge complex networks has become particularly demanding in computational terms. A task of paramount importance for understanding the structure of social networks as well as of many other real-world systems is to identify *communities*, that is, sets of nodes that are more densely connected to each other than to other nodes of the network. In this paper we propose two algorithms for community detection in networks, by employing the *neighborhood overlap* metric and appropriate spanning tree computations.

**Keywords:** Community detection · Neighborhood overlap · Hierarchical clustering · Edge betweenness · Modularity · Social networks · Spanning trees

## 1 Introduction

Over the last few decades, advances in technology and the rise of the Internet have led to numerous online social networks, like Facebook, Twitter, LinkedIn, and Instagram, where people interact and exchange information at an unprecedented rate forming a plethora of virtual groups, communities and societies. Apart from its own interest, the study and analysis of social networks finds applications on complex networks that appear in various other scientific fields. Scientists working on different disciplines like sociology, computer science, anthropology, psychology, biology, and physics are interested in the discovery of various structural and statistical properties that characterize complex networks [2]. One of the most important problems in analyzing such networks is the detection of

---

The order of authors is alphabetical; each author had an equal contribution to this work.

© Springer Nature Switzerland AG 2019
Y. Disser and V. S. Verykios (Eds.): ALGOCLOUD 2018, LNCS 11409, pp. 13–24, 2019.
https://doi.org/10.1007/978-3-030-19759-9_2

communities based on observable connections and interaction among users or components of the network. Prediction of human emotions, influence propagation, sentiment analysis, opinion dynamics, protein interaction are some of the ever-expanding fields for which community detection is highly relevant.

In this work we are proposing two new algorithms for community detection in networks that can be represented by unweighted graphs, that is, networks in which only information regarding connections between parts of the network is available. Our algorithms are hierarchical clustering methods, that make novel use of the Neighborhood overlap (nover) metric and spanning tree computations. We compare our algorithms with two related well-known algorithms Louvain and Girvan-Newmann (GN) by performing experiments on random graphs as well as on real-world networks. The contribution of this work is twofold: first, we manage to obtain a fast parallelizable algorithm based on spanning tree computations and second we reveal cases where the use of the nover similarity measure can enhance community detection.

## 1.1   Community Detection

A community in a network is a collection of nodes that are more densely connected to each other than to nodes outside the community. Detecting communities thus helps us identify nodes with common preferences, properties or behavior, unveil interactions and evaluate relationships among them and often discover hidden information.

Currently, there are quite a few methods and techniques that deal with finding communities. As an example, a lot of techniques identify edges that link different communities. In order to find such edges, various centrality measures, such as node or edge betweenness, are used. Popular approaches attempt to discover a hierarchical structure in a network, and create communities that maximize or minimize some evaluation function. Well-known community detection algorithms are the Girvan-Newman algorithm, which is based on the *edge betweenness* metric [8], the Louvain algorithm [4], and the Label propagation algorithm [14], to name only a few.

Often, edges within the same communities tend to have lower traffic, in case of information or other flow among nodes, a fact that is reflected to smaller edge betweenness (see below) compared to that of edges belonging to different communities. Thus removing edges with high edge betweenness seems a reasonable approach in order to partition the network into communities. While doing so, it is usual to keep track of the quality of the formed partitions using a metric called *modularity*, which is a well established community quality measure. A third measure of interest in this work is the *neighborhood overlap* which reveals the strength of bonds between a pair of nodes in terms of the fraction of neighbors that are common to both. These three measures are described in more detail below.

## 1.2   Terminology

For a graph $G(V, E)$, which models a network, where $V$ is the set of nodes (users), and $E$ is the set of edges (connections between nodes), we define the following notions and measures.

**Edge Betweenness.** Edge betweenness (eb) of an edge $e \in E$ defines how important that edge is with respect to shortest paths that connect each pair of nodes in $G$. More specifically, eb$(e)$ is defined as the sum, over all pairs of nodes $i, j$, of the ratio of the number of shortest paths between $i$ and $j$ passing through edge $e$ over the total number of shortest paths between $i$ and $j$. It is based on the assumption that if much of the traffic of a network passes through an edge (assuming that traffic is routed through shortest paths) then that edge is likely to connect different communities.

**Neighborhood Overlap.** The neighborhood overlap (nover) of an edge $e = (u, v)$ is a measure of embeddedness of $e$ defined as the ratio of the number of common neighbors of $u$ and $v$ to the number of nodes that are neighbors of either $u$ or $v$:

$$\text{nover}(u, v) = \frac{|N_u \cap N_v|}{|N_u \cup N_v|}, \tag{1}$$

where $N_u$ denotes the set of neighbors of node $u$.

When an edge $e$ is a local bridge, then nover$(e) = 0$, and edges with very small nover value can be seen as almost local bridges. An edge with low nover score is considered a weak tie and an edge with high nover score is a strong tie. Note that nover$(u, v)$ is in fact the Jaccard index of the two neighborhood sets $N_u, N_v$ and measures the nodes similarity.

**Modularity.** One way to measure the quality of the formed community structure is the modularity [13]. Modularity $Q$ is a scalar value, $-1 \leq Q \leq 1$, and it measures the density of the nodes within the same community compared to a random assignment of edges. The larger the modularity score, the better the partitioning of the nodes into communities. It is used to compare the communities obtained by different methods. It is calculated as,

$$Q = \frac{1}{2m} \cdot \sum_{i,j} \left[ A_{ij} - \frac{k_i k_j}{2m} \right] \cdot \delta(c_i, c_j) \tag{2}$$

where $m$ is the number of edges, $A_{ij}$ is the weight of an edge between nodes $i$ and $j$, $k_i$ is the degree of node $i$, $c_i$ is the community to which node $i$ belongs to, and $\delta$ is a function such that $\delta(c, c') = 1$ if $c = c'$ else 0. A modularity value close to 0 or negative indicates low community structure, while a value well above 0 indicates high community structure.

## 1.3   Related Work

Various approaches have been proposed in recent years to solve the community detection problem. Most popular among these are: (a) optimization methods, which aim to maximize or minimize an objective function, and (b) hierarchical methods that are either divisive or agglomerative.

In the seminal paper of Girvan and Newman [8] they define the eb centrality measure and propose the GN algorithm that uses this measure. GN iteratively removes edges of higher eb centrality, thus forming connected components that correspond to communities. The main disadvantage of this algorithm is that it is computationally expensive (since it recomputes the eb values for all edges in each step) and thus not scalable. The running time in the worst case is $O(|E|^2|V|)$.

A two phase algorithm for weighted graphs was proposed in [4], known as the Louvain algorithm, that runs in $O(|E|)$ time. In the first phase, for each node it iteratively calculates the modularity obtained by including the node to the community of each of its neighbors, and then places this node into the community that gives the highest modularity. In the second phase, it creates a meta-graph in which communities are represented as meta-nodes and self-loops represent edges internal to the communities. The two phases are repeated on the meta-graph. This algorithm has a tendency to overlook small communities. In general, methods that use the modularity metric to optimize the community detection are known to suffer from the *resolution limit* effect [7], which refers to the fact that communities smaller than some threshold may not be discovered. Furthermore, the Louvain algorithm cannot efficiently explore the hierarchical structure of the network (if such a structure is present).

The main idea in [12] is to use the nover score to differentiate weak from strong ties. The nover scores are stored in a minimum heap and all edges with a score smaller than a threshold value are considered weak ties and are removed. The problem with this approach is correctly deciding the threshold value. Larger values of the threshold value could disintegrate the communities and smaller values could make two different communities merge.

Regarding other successful methods for community detection in the setting where overlapping communities are also sought, BigClam [16] should be mentioned, a method that uses matrix factorization in order to discover overlapping and non overlapping communities in large scale networks, and another approach by Ahn, Bagrow and Lehmann [2] that also discovers overlapping communities by partitioning edges instead of nodes. Both these approaches work on the global structure of the network. In a different direction, a hierarchical scalable edge clustering algorithm presented in [9] aims at discovering overlapping local communities of a seed node; a similar approach when the graph is given as a stream is described in [10]. The latter manages to maintain minimal information about the whole graph and the formed communities, thus using space sublinear in the number of edges.

## 1.4    Our Contribution

In order to propose scalable and parallelizable algorithms that build on the importance of the nover score we introduce two approaches. Our first approach modifies Louvain algorithm by assigning weights equal to their nover score; recall that Louvain is designed to work on weighted graphs. We run several experiments in order to estimate the importance of this modification. The obtained results in most cases show that the modified algorithm performs better, therefore nover values seem worth taking into account.

Our second approach uses the eb centrality in conjunction with the nover metric in order to remove edges and form connected components (communities). The eb metric is also used by the GN algorithm; however, in contrast to the GN approach, we start by computing a maximum spanning tree using the nover weights, thus considerably reducing the eb computations which take place on the tree only. This yields a faster and parallelizable approach. It turns out that the use of a spanning tree together with nover scores helps sparsify a graph without substantial loss of information. In particular, the experimental results show that while the time gain of our ST algorithm compared to GN is important, the modularity values are of the same order of magnitude and in some cases even exceed those of the GN algorithm. Moreover, the high time requirements of the GN algorithm render it inapplicable to larger networks, while ST does not seem to suffer from similar limitations.

## 2    Neighborhood Overlap-Based Approaches

We consider unweighted, undirected graphs. In both our algorithms, we use the nover score of edges Eq. (1) in order to assign weights to them. Intuitively, we want to differentiate weak from strong ties. We consider strong ties as more probable to connect nodes within the same community. Let us assume $G(V, E)$ is an input graph. The first step of both our algorithms is to calculate the nover score for every edge $e \in E$. Thus, we get a weighted graph.

In Algorithm 1 (nover-Louvain), this preprocessing is the only modification with respect to the original Louvain algorithm. Note that this increases by at most an $O(\Delta)$ factor the time complexity of the algorithm, where $\Delta$ is the maximum degree of the network. This is because we need $O(|E|\Delta)$ time for computing the nover of all edges, since computing the common neighbors of an edge can be done in $O(\Delta)$ time. Combining with the $O(|E|)$ complexity of the original Louvain we get a total time complexity of $O(|E|\Delta)$.

In Algorithm 2 (ST), in the first phase we make use of the nover edge weights in order to perform the maximum spanning tree computation. In the second phase, we calculate the eb score of each edge, taken over the constructed spanning tree. We then repeatedly remove one edge at a time, in non-increasing order of their eb score; thus, in each repetition we increase the number of communities by one. Since the optimal number of communities is not known beforehand we keep repeating until all edges are removed; the output is the set of communities $C$ that yields the highest modularity.

**Algorithm 1.** Modified Louvain community detection by neighborhood overlap (nover-Louvain).

---

**Input:** $G(V, E)$
**Output:** Set of communities $C$ of maximum modularity $Q$
  **for each** edge $e = (u, v) \in E$ **do**
    $\text{nover}(e) = |N_u \cap N_v| \, / \, |N_u \cup N_v|$
    $w(e) \leftarrow \text{nover}(e)$
  **end for**
  $C \leftarrow \text{Louvain}(G, w)$
  **return** $C$

---

The time complexity of Algorithm 2 is analyzed as follows:

– $O(|E|\Delta)$, for computing the nover of all edges of the graph, as explained above
– $O(|E| + |V| \log |V|)$, for computing the maximum spanning tree using Prim's algorithm
– $O(|V|)$, for computing the eb values for all edges in the spanning tree
– $O(|V| \log |V|)$ for sorting the eb values of the tree edges
– $O(|E||V|)$, for the *while* loop since $O(|E|)$ suffices in order to compute the communities and modularity in each iteration.

Therefore, an upper bound on the total running time is $O(|E||V|)$. Note that this may be slightly improved by a more tight analysis of the *while* loop, but such an analysis would not improve the worst case bound.

# 3    Experimental Results

We evaluate our methods using experiments on various datasets and compare them against the most established algorithms, namely the Louvain[1] algorithm [4] and the Girvan-Newman (GN) algorithm [8]. These two algorithms were already implemented in the `python igraph` library so we have chosen to implement our new algorithms nover-Louvain and ST using the same library. We focus on two criteria for the comparison of the algorithms, namely the modularity and the number of communities found.

We have experimented with two types of networks and corresponding datasets: synthetic and real-world data. Regarding synthetic datasets we employed two different random graph models, namely the Barabási-Albert model and the Erdős-Rényi model; as for real-world datasets we considered the Zachary's karate club network, and the Facebook network. Our results are summarized in Tables 1, 2, 3 and are discussed below.

---

[1] Note that the plain Louvain algorithm, can be applied on unweighted graphs by setting all edge weights equal to 1.

---

**Algorithm 2.** Community detection by neighborhood overlap and maximum spanning tree (ST).

---

**Input:** $G(V, E)$
**Output:** Set of communities $C$ with maximum modularity $Q$
  **for each** edge $e = (u, v) \in E$ **do**
    $\mathsf{nover}(e) = |N_u \cap N_v| \, / \, |N_u \cup N_v|$
    $w(e) \leftarrow \mathsf{nover}(e)$
  **end for**
  $G'(V, E') \leftarrow$ calculate Maximum Spanning Tree$(G, w)$
  **for each** $e \in E'$ **do**
    $\mathsf{eb}(e) \leftarrow$ calculate Edge Betweenness on $e$
  **end for**
  Initialize $C \leftarrow \{V\}$, $Q \leftarrow$ modularity of $C$ in $G(V, E)$       ▷ one community
  Sort all edges in $E'$ in non-increasing order of $\mathsf{eb}(e)$
  **while** $E'$ is nonempty **do**
    Remove the edge $e$ of highest $\mathsf{eb}(e)$ from $E'$   ▷ next edge in sorted list of edges
    $C' \leftarrow$ community structure implied by $E'$ ▷ set of components, partitioning $V$
    $Q' \leftarrow$ modularity of $C'$ in $G(V, E)$     ▷ modularity is wrt the original graph
    **if** $Q' > Q$ **then**
      $Q \leftarrow Q'$
      $C \leftarrow C'$
    **end if**
  **end while**
  **return** $C$

---

### 3.1 Synthetic Network Models

**Barabási-Albert Graphs.** We use the Barabási-Albert [3] synthetic random graph model, which is a well known model for generating random networks. These networks have a power-law distribution property. A power-law distribution property implies that only few nodes in a network have a high degree, a property that is common in social networks and the Internet.

We use different values on parameters for our experiments in order to find out in which cases our new algorithms behave better than the existing ones and when the results are getting worse. The number of nodes varies from 40 to 1000, while the number of edges vary from 114 to 1997. The dataset generator actually does not permit to determine the number of edges but only the minimum degree. The minimum degree value in our experiments is either 2 or 3. Since the graphs resulting from the given parameters are random, the obtained results sometimes vary considerably (see Table 1). We have chosen to present two experiments for each parameter setting, the most extreme ones with respect to the best modularity observed among experiments of the same setting.

A remarkable case is that of graphs with 150 nodes, 297 edges and minimum node degree 2, in which the Louvain algorithm achieves modularity 0.139 for a certain graph and 0.457 for another one, both graphs having been obtained by using the same parameters. The nover-Louvain algorithm achieves, in both graphs, better results; the difference in modularity in the same graphs is again remarkable (0.189 and 0.459, respectively). For the GN algorithm this difference is even larger compared to the other algorithms. Our ST algorithm on the contrary produces results close to those of Louvain and nover-Louvain; in fact it outperforms Louvain for the first of the two graphs.

In the case of networks consisting of 400 nodes, 1194 edges and minimum degree 2, the results of our new algorithms are encouraging. In almost all cases, nover-Louvain outperforms all other algorithms and the results of ST are comparable to the GN results, which is quite interesting if we take into consideration the much better running time of ST compared to GN.

**Table 1.** Results for Barabási-Albert graphs (mod. = modularity, #cl. = number of clusters).

| #nodes | deg | #edges | Louvain [4] | | nover-Louvain | | GN [8] | | ST | |
|---|---|---|---|---|---|---|---|---|---|---|
| | | | mod. | #cl. | mod. | #cl. | mod. | #cl. | mod. | #cl. |
| 40 | 3 | 114 | 0.245 | 6 | 0.254 | 5 | 0.133 | 9 | 0.198 | 3 |
| 40 | 3 | 114 | 0.25 | 6 | 0.249 | 5 | 0.088 | 7 | 0.218 | 7 |
| 80 | 3 | 234 | 0.282 | 8 | 0.269 | 6 | 0.167 | 32 | 0.255 | 7 |
| 80 | 3 | 234 | 0.289 | 7 | 0.327 | 7 | 0.224 | 21 | 0.263 | 4 |
| 100 | 2 | 197 | 0.43 | 9 | 0.448 | 9 | 0.42 | 10 | 0.41 | 7 |
| 100 | 2 | 197 | 0.41 | 7 | 0.446 | 9 | 0.373 | 15 | 0.375 | 8 |
| 150 | 2 | 297 | 0.448 | 8 | 0.457 | 9 | 0.42 | 11 | 0.401 | 9 |
| 150 | 2 | 297 | 0.453 | 8 | 0.456 | 9 | 0.44 | 11 | 0.413 | 10 |
| 150 | 3 | 444 | 0.139 | 9 | 0.189 | 6 | 0.026 | 10 | 0.147 | 4 |
| 150 | 3 | 444 | 0.457 | 9 | 0.459 | 9 | 0.43 | 13 | 0.409 | 9 |
| 200 | 3 | 594 | 0.339 | 9 | 0.333 | 9 | 0.254 | 41 | 0.267 | 6 |
| 200 | 3 | 594 | 0.334 | 8 | 0.334 | 8 | 0.257 | 40 | 0.281 | 8 |
| 400 | 2 | 1194 | 0.476 | 14 | 0.486 | 16 | 0.446 | 23 | 0.427 | 11 |
| 400 | 2 | 1194 | 0.487 | 12 | 0.489 | 13 | 0.441 | 27 | 0.409 | 6 |
| 400 | 3 | 1194 | 0.358 | 12 | 0.36 | 13 | 0.297 | 44 | 0.284 | 7 |
| 400 | 3 | 1194 | 0.353 | 12 | 0.364 | 12 | 0.287 | 57 | 0.296 | 9 |
| 1000 | 2 | 1997 | 0.511 | 19 | 0.51 | 20 | 0.467 | 51 | 0.404 | 7 |
| 1000 | 2 | 1997 | 0.511 | 19 | 0.513 | 20 | 0.479 | 50 | 0.43 | 9 |

We observe that the modified nover-Louvain gives slightly better results compared to Louvain in almost all cases. This leads us to conclude that the nover similarity is an important measure which should be further taken into consideration for community detection purposes.

**Erdős-Rényi Graphs.** In this random graph model [6], a random graph $G(n, p)$ is constructed by connecting $n$ nodes randomly. Each edge exists in $G$ with probability $p$, independent from every other edge. The average degree is $np$.

**Table 2.** Results for Erdős-Rényi graphs (mod. = modularity, #cl. = number of clusters).

| #nodes | prob | Louvain [4] | | nover-Louvain | | GN [8] | | ST | |
|--------|------|------|------|------|------|------|------|------|------|
| | | mod. | #cl. | mod. | #cl. | mod. | #cl. | mod. | #cl. |
| 40 | 0.1 | 0.363 | 6 | 0.378 | 5 | 0.358 | 8 | 0.327 | 6 |
| 40 | 0.1 | 0.363 | 6 | 0.415 | 6 | 0.298 | 11 | 0.323 | 7 |
| 80 | 0.1 | 0.273 | 6 | 0.277 | 7 | 0.199 | 25 | 0.24 | 5 |
| 80 | 0.1 | 0.249 | 8 | 0.271 | 6 | 0.172 | 26 | 0.217 | 5 |
| 120 | 0.1 | 0.23 | 8 | 0.233 | 8 | 0.123 | 54 | 0.171 | 11 |
| 120 | 0.1 | 0.233 | 8 | 0.234 | 6 | 0.098 | 67 | 0.196 | 8 |
| 200 | 0.1 | 0.172 | 9 | 0.175 | 8 | 0.05 | 105 | 0.122 | 7 |
| 200 | 0.1 | 0.185 | 7 | 0.178 | 8 | 0.054 | 109 | 0.133 | 11 |

The graphs we created have 40 to 200 nodes and the probability of the existence of an edge is 0.1. The Erdős-Rényi graphs follow the Poisson degree distribution and can be used for evaluating community detection algorithms [15].

In our experiments in Table 2, one can see that as the number of nodes increases, keeping the same edge probability, the modularity (columns labeled 'mod.') tends to decrease steadily and the number of clusters (columns labeled '#cl.') tends to increase with the same pace for all algorithms, except for GN. For GN the modularity decreases faster and the number of communities increments a lot, and one can assume that a lot of smaller communities are formed. A possible explanation is that as the average degree increments, the GN algorithm will have to remove more edges, evenly distributed through the network, in order to get the partitions. On the other hand, Louvain and nover-Louvain tend to create larger communities since nodes exhibit the preferential attachment behavior. The ST algorithm also creates larger communities by providing a more robust initial structure through the nover scores and the spanning tree computation.

## 3.2   Real-World Networks

Synthetic datasets allow us to work with randomly generated graphs in which the structure is determined by certain parameters. On the other hand, it makes sense to evaluate our algorithms' performance in real datasets where we know in advance the community structure in some cases. We have thus chosen to experiment with the Zachary's karate club network and one large network, namely Facebook.

**Zachary's Karate Club.** This is a famous network model, due to Zachary's observations, explained in his study [18]. Over the course of two years, he monitored members of a karate club and relationship among those members. Over time, dispute arose between team's instructor and administrator which resulted in club splitting in two separate clubs. Half of the original club members joined the new club [8]. Zachary built a network model based on his observations, which included 34 nodes and 78 links.

For Zachary's dataset we get some interesting results. When we compare the algorithms in respect to the modularity value as shown in Table 3, our algorithms do not perform as good as Louvain or GN. However, when we compare them regarding the number of clusters, ST performs much better. It correctly found the 2 communities (Fig. 1a) as in the paper and a small one containing just two nodes (node 8 and 30 in Zachary's paper [18]). This is very close to the split of nodes that actually took place in the Karate Club network. This example provides evidence that modularity, although a generally accepted measure for evaluating a proposed community partition, may fail to correctly reflect the actual community structure. On the other hand, GN finds 5 communities (Fig. 1b), while both nover-Louvain and Louvain find 4 communities (Figs. 1c, d).

**Facebook Friendship Network.** This is a friendship network between Facebook users [1,11]. In this network a node represents a user and an edge indicates

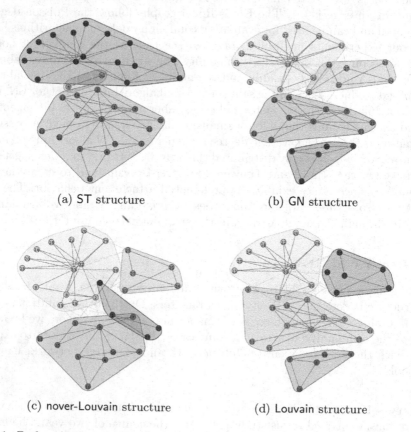

(a) ST structure                    (b) GN structure

(c) nover-Louvain structure         (d) Louvain structure

**Fig. 1.** Zachary's Karate club: community structure as obtained by ST, GN, nover-Louvain, and Louvain algorithms.

friendship between the corresponding users. This is an unweighted directed network with 2888 nodes and 2981 edges. For the purpose of this study we have ignored the edge directions. It is clear that this network is sparse, hence some of the nodes are too far from each other.

For larger graphs it makes sense to focus on parallelizable algorithms, appropriate for distributed implementation. We thus experimented with the Facebook network using GN and ST algorithms, since the Louvain algorithm in its original form seems to be inherently sequential. Our results show that ST yields a modularity very close to the one obtained by GN (see Table 3); since ST is a much faster, parallelizable algorithm it provides a possibly useful alternative whenever time gains and parallelization are more important than a slight modularity decrease.

**Table 3.** Results for real-world datasets.

| Dataset | #nodes | #edges | GN [8] | | ST | |
|---------|--------|--------|--------|-----------|--------|-----------|
| | | | modularity | #clusters | modularity | #clusters |
| Zachary's karate club | 34 | 78 | 0.401 | 5 | 0.372 | 3 |
| Facebook | 2888 | 2981 | 0.807 | 8 | 0.804 | 7 |

# 4  Conclusions

The volume of available data and the size of networks of today in various fields, such as social and biological networks, call for efficient distributed methods to identify communities in large scale networks.

Towards this goal we first introduced the ST algorithm, that uses the nover score to assign weights on edges in order to compute a spanning tree and then uses the eb centrality to split the tree into a forest. Each tree in the forest yields a community. The main advantage of the ST algorithm, is that by using nover and the spanning tree it renders computations less time consuming (without losing much on the quality of the results) compared to the GN algorithm. Note also that the ST algorithm is parallelizable and thus is appropriate for distributed implementation.

We next explored the idea that the nover score, being a measure of the similarity of two neighbor nodes, may improve existing algorithms if used in a preprocessing phase. We thus obtained a second algorithm by adding such a preprocessing to the Louvain algorithm. The new algorithm nover-Louvain indeed performed better in several cases.

As future work, we plan to compare these algorithms using different objective criteria to evaluate the communities formed. As mentioned earlier, modularity is a measure of the quality of the communities that may have some drawbacks, as indicated by the Zachary's karate club example. Thus we intend to explore additional measures, such as the Normalized Mutual Information (NMI) score

[5] and the conductance [17] in order to compare the different approaches. Furthermore, we also plan to experiment with alternative graph metrics to obtain edge weights for the preprocessing phase, in order to see if they can lead to further improvement of the proposed algorithms nover-Louvain and ST.

# References

1. Facebook (nips) network dataset - KONECT, April 2017. http://konect.uni-koblenz.de/networks/ego-facebook
2. Ahn, Y.Y., Bagrow, J.P., Lehmann, S.: Link communities reveal multiscale complexity in networks. Nature **466**(9), 761–764 (2010)
3. Barabási, A.L., Albert, R.: Emergence of scaling in random networks. Science **286**(5439), 509–512 (1999)
4. Blondel, V.D., Guillaume, J.L., Lambiotte, R., Lefebvre, E.: Fast unfolding of communities in large networks. J. Stat. Mech. Theory Exp. **2008**(10), P10008 (2008)
5. Danon, L., Díaz-Guilera, A., Duch, J., Arenas, A.: Comparing community structure identification. J. Stat. Mech. Theory Exp. **2005**(09), 09008 (2005)
6. Erdös, P., Rényi, A.: On random graphs I. Publicationes Mathematicae Debrecen **6**, 290 (1959)
7. Fortunato, S., Barthelemy, M.: Resolution limit in community detection. Proc. Nat. Acad. Sci. **104**(1), 36–41 (2007)
8. Girvan, M., Newman, M.E.: Community structure in social and biological networks. Proc. Nat. Acad. Sci. **99**(12), 7821–7826 (2002)
9. Liakos, P., Ntoulas, A., Delis, A.: Scalable link community detection: a local dispersion-aware approach. In: IEEE International Conference on Big Data, pp. 716–725 (2016)
10. Liakos, P., Ntoulas, A., Delis, A.: COEUS: community detection via seed-set expansion on graph streams. In: IEEE International Conference on Big Data, pp. 676–685 (2017)
11. McAuley, J., Leskovec, J.: Learning to discover social circles in ego networks. In: Advances in Neural Information Processing Systems, pp. 548–556 (2012)
12. Meghanathan, N.: A greedy algorithm for neighborhood overlap-based community detection. Algorithms **9**(1), 8 (2016)
13. Newman, M.E., Girvan, M.: Finding and evaluating community structure in networks. Phys. Rev. E **69**(2), 026113 (2004)
14. Raghavan, U.N., Albert, R., Kumara, S.: Near linear time algorithm to detect community structures in large-scale networks. Phys. Rev. E **76**(3), 036106 (2007)
15. Reichardt, J., Bornholdt, S.: Statistical mechanics of community detection. Phys. Rev. E **74**, 016110 (2006)
16. Yang, J., Leskovec, J.: Overlapping community detection at scale: a nonnegative matrix factorization approach. In: 6th ACM WSDM 2013, pp. 587–596 (2013)
17. Yin, H., Benson, A.R., Leskovec, J., Gleich, D.F.: Local higher-order graph clustering. In: Proceedings of the 23rd ACM SIGKDD, pp. 555–564 (2017)
18. Zachary, W.W.: An information flow model for conflict and fission in small groups. J. Anthropol. Res. **33**(4), 452–473 (1977)

# Colocation, Colocation, Colocation: Optimizing Placement in the Hybrid Cloud

Srinivas Aiyar[1], Karan Gupta[1], Rajmohan Rajaraman[2], Bochao Shen[2], Zhifeng Sun[2], and Ravi Sundaram[2(✉)]

[1] Nutanix, Inc., San Jose, CA, USA
{sriniva.aiyar,karan.gupta}@nutanix.com
[2] Northeastern University, Boston, MA, USA
{rraj,ordinary,austin,koods}@ccs.neu.edu

**Abstract.** Today's enterprise customer has to decide how to distribute her services among multiple clouds - between on-premise private clouds and public clouds - so as to optimize different objectives, e.g., minimizing bottleneck resource usage, maintenance downtime, bandwidth usage or privacy leakage. These use cases motivate a general formulation, the *uncapacitated* (A defining feature of clouds is their *elasticity* or ability to scale with load) multidimensional load assignment problem - VITA(F) (Vectors-In-Total Assignment): the input consists of $n$, $d$-dimensional load vectors $\bar{V} = \{\bar{V}_i | 1 \leq i \leq n\}$, $m$ cloud buckets $B = \{B_j | 1 \leq j \leq m\}$ with associated weights $w_j$ and assignment constraints represented by a bipartite graph $G = (\bar{V} \cup B, E \subseteq \bar{V} \times B)$ restricting load $\bar{V}_i$ to be assigned only to buckets $B_j$ with which it shares an edge (In a slight abuse of notation, we let $B_j$ also denote the subset of vectors assigned to bucket $B_j$). $F$ can be any operator mapping a vector to a scalar, e.g., max, min, etc. The objective is to partition the vectors among the buckets, respecting assignment constraints, so as to achieve

$$\min[\sum_j w_j * F(\sum_{\bar{V}_i \in B_j} \bar{V}_i)]$$

We characterize the complexity of VITA(min), VITA(max), VITA(max − min) and VITA($2^{nd}$ max) by providing hardness results and approximation algorithms, *LP-Approx* involving clever rounding of carefully crafted linear programs. Employing real-world traces from Nutanix, a leading hybrid cloud provider, we perform a comprehensive comparative evaluation versus three natural heuristics - *Conservative*, *Greedy* and *Local-Search*. Our main finding is that on real-world workloads too, *LP-Approx* outperforms the heuristics, in terms of quality, in all but one case.

## 1 Introduction

The launch of EC2 in 2006 by AWS [1] heralded the explosive growth in cloud computing. Cloud computing is an umbrella term for computing as an utility.

© Springer Nature Switzerland AG 2019
Y. Disser and V. S. Verykios (Eds.): ALGOCLOUD 2018, LNCS 11409, pp. 25–45, 2019.
https://doi.org/10.1007/978-3-030-19759-9_3

It enables $24 \times 7$ Internet-based access to shared pools of configurable system resources and real-time provision-able higher-level services. Public clouds enable organizations to focus on their core businesses instead of spending time and money on IT infrastructure and maintenance. One of the major benefits of clouds is that they are *elastic*[1] (which we model in this paper as uncapacitated). This allows enterprises to get their applications[2] up and running quicker, and rapidly adjust resources to meet fluctuating and unpredictable business demand.

Today, in addition to AWS, Microsoft's Azure [5] and the Google Cloud [3] are the other major public cloud platforms. But the advent of multiple clouds means that enterprises are faced with several new questions, of which the following are some examples: How much of their load should they keep on-premise and how much should they colocate (or place) in public clouds? How should they mix and match the various options to save money without sacrificing customer satisfaction? A number of enterprise software companies such as HPE [4] and startups such as Turbonomic [7], Datadog [2] and RightScale [6] are beginning to provide software and service solutions to these problems.

At the same time this is also a fertile area for new problems with the potential for clever theoretical solutions to have practical impact. In this paper we provide a framework - VITA: Vectors-In-Total Assignment - that captures a variety of interesting problems in the area of hybrid clouds with interesting theoretical challenges. In the subsection that follows we list a few typical use cases captured by the VITA framework.

## 1.1  Motivation and Model

**Scenario 1. Minimizing Peak Pricing:** Consider an enterprise customer that has a choice of several different cloud providers at which to host their VMs (virtual machines). The requirements of each VM can be characterized along several different resource dimensions such as compute (CPU), network (latency, bandwidth), storage (memory, disk) and energy. When different virtual machines are placed in the same elastic resource pool (cloud), their load across each dimension is accrued additively (though, of course the different dimensions can be scaled suitably to make them comparable). A typical pricing contract will charge based on the most bottle-necked dimension since peak provisioning is the biggest and most expensive challenge for the resource provider. And different providers may have different rates based on differing infrastructure and their cost for installation and maintenance. The natural question then arises - what is the optimal way for the enterprise customer to distribute the load amongst the different cloud providers so as to minimize total cost?

---

[1] Elastic usually means that clouds can be considered to have infinite capacity for the operating range of their customers. In this paper we ignore fine-grained time-based definitions such as in [20].

[2] In the scope of this paper *application* refers to a collection of VMs and containers working in concert.

**Scenario 2. Minimizing Maintenance Downtime:**Hosts and services, (and occasionally even data centers) need to be powered down every so often for maintenance purposes, e.g. upgrading the software version (or installing a new HVAC system in a data center). Given this reality, how should the application (collection of virtual machines and/or containers collectively performing a task or service), be allocated to the different hosts so as to minimize the aggregate disruption? This scenario also applies to industrial machines where different factories (or floors of a factory) need to be shut down for periodical maintenance work.

**Scenario 3. Preserving Privacy:** Consider a set of end-users each with its own (hourly) traffic profile accessing an application. We wish to partition the application components across a set of clouds such that by observing the (hourly volume of) traffic flow of any single cloud it is not possible to infer which components are colocated there. This leads to the following question - how should we distribute load across clouds in order to minimize the maximum hourly variation in aggregate traffic? As an analogy, the situation here is similar to the problem of grouping households such that the variation of energy usage of a group is minimized making it difficult for thieves to infer who has gone on vacation.

**Scenario 4. Burstable Billing:** Most Tier 1 Internet Service Providers (ISPs) use burstable billing for measuring bandwidth based on peak usage. The typical practice is to measure bandwidth usage at regular intervals (say 5 min) and then use the 95th percentile as a measure of the sustained flow for which to charge. The 95th percentile method more closely reflects the needed capacity of the link in question than tracking by other methods such as mean or maximum rate. The bytes that make up the packets themselves do not actually cost money, but the link and the infrastructure on either end of the link cost money to set up and support. The top 5% of samples are ignored as they are considered to represent transient bursts. Burstable billing is commonly used in peering arrangements between corporate networks. What is the optimal way to distribute load among a collection of clouds, public and private, so as to minimize the aggregate bandwidth bill?

The above scenarios constitute representative situations captured by the *uncapacitated* multidimensional load assignment problem framework - VITA. A host of related problems from a variety of contexts can be abstracted and modeled as VITA(F): the input consists of $n$, $d$-dimensional load vectors $\bar{V} = \{\bar{V}_i | 1 \le i \le n\}$ and $m$ cloud buckets $B = \{B_j | 1 \le j \le m\}$ with associated weights $w_j$ and assignment constraints represented by a bipartite graph $G = (\bar{V} \cup B, E \subseteq \bar{V} \times B)$ that restricts load $\bar{V}_i$ to be assigned only to those buckets $B_j$ with which it shares an edge. Here, $F$ can be any operator mapping a vector to a scalar, such as projection operators, max, min, etc. Then the goal is to partition the vectors among the buckets, respecting the assignment constraints, so as to minimize

$$\sum_j w_j * F(\sum_{\bar{V}_i \in B_j} \bar{V}_i)$$

where, in a slight abuse of notation, we let $B_j$ also denote the subset of vectors assigned to bucket $B_j$. VITA stands for Vectors-In-Total Assignment capturing the problem essence - vectors assigned to each bucket are totaled. Unless otherwise specified we use $i$ to index the load vectors, $j$ to index the cloud buckets and $k$ to index the dimension. We let $\bar{V}_i(k)$ denote the value in the $k$'th position of the vector $\bar{V}_i$.

We now explain how VITA(F) captures the aforementioned scenarios. In general, dimensions will either represent categorical entities such as resources (e.g., CPU, I/O, storage, etc.,) or time periods (e.g., hours of the day or 5-min intervals, etc.,). We gently remind the reader to note that in each of the scenarios the elasticity of the clouds is a critical ingredient so that contention between vectors is not the issue. The set of scenarios we present are but a small sample to showcase the versatility and wide applicability of the VITA framework.

Scenario 1 is captured by having a vector for each VM, with each dimension representing its resource requirement[3]; constraints representing placement or affinity requirements [21], weights $w_j$ representing the rates at different cloud providers. Then minimizing the sum of prices paid for peak resource usage at each cloud is just the problem VITA(max).

In Scenario 2 each dimension represents the resource (say, CPU utilization) consumed by the application in a given time period, e.g. the vector for an application could have 24 dimensions one for each hour in the day. Once the application is assigned to a data center (or cloud or cluster) it is clear that disruption is minimized if the maintenance downtime is scheduled in that hour where total resource utilization is minimum. Then minimizing the aggregate disruption is captured by the problem VITA(min).

The dimensions in Scenario 3 are the hours of the day and the resource in question is the traffic. To prevent leakage of privacy through traffic analysis the goal is to distribute the application components across clouds so that the range between the peak and trough of traffic minimized. This problem is exactly represented as VITA(max − min).

In Scenario 4, we have vectors for each application with 20 dimensions one for each 5th percentile [28, 29] or ventile of the billing period[4]. Then minimizing the aggregate bandwidth bill under the burstable, or 95th percentile, billing method is VITA($2^{nd}$ max).

## 1.2   Our Results

All the problems we consider are in NP [18]. For VITA(min) and VITA(max) we present our results as a lattice - see Figs. 1 and 2. For any given F, VITA(F) can be partitioned into a lattice of 4 different problem spaces based on the following 2 criteria: 1. constraints, and 2. dimensionality. The 4 different problem spaces arise from the Cartesian product: {unconstrained, constrained} X

---

[3] For time-varying requirements the problem can be modeled by #resources x #time-periods dimensions.

[4] This is a modeling approximation and does not exactly capture 5 min samples.

{bounded, unbounded}. *Unconstrained* refers to the situation where there is no bipartite graph representing constraints, i.e. any load vector may be placed in any bucket. And, *Bounded* refers to the situation where each load vector has a fixed dimension (independent of $n$). It should be clear that the simplest of the 4 spaces is unconstrained, bounded VITA(F) and the most general is the constrained, unbounded version of VITA(F). We present our results, algorithms and hardness, for the different F, in the form of a lattice. In each of the figures, the algorithmic results are displayed only at the highest possible node in the lattice, since it automatically applies to all nodes in the downward-closure; similarly, hardness results are presented at the lowest possible node since they apply to all nodes in the upward-closure. Further, our hardness results use only uniform weights whereas our algorithmic results work for general weights.

Our theory results are as follows:

- VITA(F) for F linear. We show that when F is linear then the problem is solvable exactly in polynomial-time. In particular VITA($avg$) is in P.
- VITA(min). Our results are summarized in Fig. 1. We show that VITA(min) is inapproximable when the dimensions are unbounded, i.e. it cannot be approximated to any finite factor. Since it is inapproximable we counter-balance this result by providing an $O(\log n, \log n)$-bicriteria approximation algorithm [25]. Our bicriteria algorithm produces an assignment of cost within $O(\log n)$ of the optimal while using no more than $O(\log n)$ copies of each bucket. The bicriteria result, which is based on rounding an LP (linear program) [27] can be considered the theoretical center-piece and contains the main ideas used in the other LP-based results in this paper.
- VITA(max). Our results are summarized in Fig. 2. Our results for VITA(max) also apply to VITA(max − min). We remind the reader that the unconstrained bounded box is empty because the algorithmic result for the harder unconstrained unbounded case (further up the lattice) applies.
- VITA($2^{nd}$ max). $2nd$ max turns out to be a particularly difficult problem from the standpoint of characterizing its computational complexity. We consider the unweighted (or uniform weights) unconstrained case and the requirement that the number of buckets exceeds the number of dimensions. With these restrictions we are able to demonstrate an LP-based approximation algorithm that achieves a logarithmic factor of approximation. We also show that unconstrained, bounded VITA($2^{nd}$ max) is weakly NP-hard [18].

This paper got its start in practical considerations at Nutanix - a leading hybrid cloud provider. Faced with a seeming plethora of different cloud colocation use-cases we wondered whether they could be tackled using a common approach. The VITA framework answers this question by providing a unified method for comparing against natural heuristics and a common basis for making pragmatic infrastructure decisions. We used real-world industrial traces from Nutanix, to conduct a detailed comparative analysis of the approximation algorithms, collectively dubbed *LP-Approx*, against 3 natural heuristics - *Conservative*, *Greedy* and *Local-Search*. *Conservative* treats each vector and its associated objective value in isolation. *Greedy* assigns vectors sequentially so as to minimize the increment in objective value. Working with a given assignment *Local-Search* swaps vectors when doing so improves the objective value. Our main finding is that from a practical standpoint too *LP-Approx* is the best in terms of solution-quality in all but one of the four cases (*Greedy* beats *LP-Approx* in the case of VITA(min)). Our work can serve as a valuable reminder of how principled and sophisticated techniques can often achieve superior quality on practical work-loads, while also providing theoretical guarantees.

 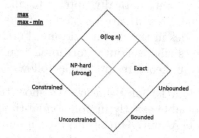

**Fig. 1.** VITA(min). The simplest unbounded case is inapproximable, and we give a bicriteria guarantee for the hardest case.

**Fig. 2.** VITA(max) and VITA(max − min). The unconstrained, cases are exactly solvable and we have tight logarithmic guarantees for the constrained unbounded case.

## 1.3   Related Work

There is extensive theory literature on multidimensional versions of schedul-ing and packing problems. [11] is an informative survey that provides a variety of new results for multidimensional generalizations of three classical packing problems: multiprocessor scheduling, bin packing, and the knapsack problem. The vector scheduling problem seeks to schedule $n$ $d$-dimensional tasks on $m$ machines such that the maximum load over all dimensions and all machines is minimized. [11] provide a PTAS for the bounded dimensionality case and poly-logarithmic approximations for the unbounded case, improving upon [22]. For the vector bin packing problem (which seeks to minimize the number of bins needed to schedule all $n$ tasks such that the maximum load on any dimension across all bins is bounded by a fixed quantity, say 1), they provide a logarithmic guarantee for the bounded dimensionality case, improving upon [32]. This result was subsequently further improved by [9]. A PTAS was provided for the multi-dimensional knapsack problem in the bounded dimension case by [17]. The key distinction between the vector scheduling problem of [11] and our framework is that they seek to minimize the maximum over the buckets and the dimensions whereas (in VITA(max)) we seek to maximize the weighted sum over buckets of the maximum dimension in each bucket. The multidimensional bin packing knapsack problems are capacitated whereas this paper deals with uncapacitated versions. There has also been a lot of work on geometric multidimensional packing where each vector is taken to represent a cuboid [10,13]. To the best of our knowledge our VITA formulation is novel - surprising given its simplicity.

There is much recent literature (in conferences such as Euro-Par, ICDCS, SIGCOMM, CCGRID, IPDPS etc.,) substantiating the motivating scenarios we provide in the introduction (Sect. 1.1) to this paper. We do not attempt to survey it in any meaningful way here. Peak provisioning and minimizing bottleneck usage is an area of active research in the systems community [12,30]. Fairness in provisioning multi-dimensional resources is studied in [19]. The use of CSP (Constraint Satisfaction Programming) in placement has been investigated [21]. Energy considerations in placement have also been explored [14–16,28,29]. Building scalable systems that provide some guarantee against traffic analysis is an area of ongoing active research [23,24,26]. Relative to the specialized literature for each use-case our treatment is less nuanced (e.g., in reality, storage is less movable than compute, services are designed for (or to give the illusion of) continuous uptime, privacy is more subtle than just defeating traffic monitoring, etc.). However, the generality of our approach enables us to abstract the essence of the different situations and apply sophisticated techniques from the theory of mathematical programming.

We present our results in the sections that follow. Section 2 presents results for linear F. Section 3 presents our results for VITA(min) while Sect. 4 contains our results for VITA(max) and VITA(max − min). VITA($2^{nd}$ max) results are presented in Sect. 5. Due to space constraints, all proofs are provided in the Appendix.

## 2    VITA(F) for Linear F

By linear F we mean one of the following two situations:

- F is a vector and $F(\bar{V}) = \bar{F} \cdot \bar{V}$ (where we abuse notation slightly and use F as a function and a vector).
- $F$ is a matrix and the weights are vectors with $*$ representing an inner-product so that $w_j * F$ is a scalar.

**Lemma 1.** *VITA(F) can be solved exactly in polynomial time for linear F.*

*Proof.* Using the linearity of F the value of the objective function can be simplified thus

$$\sum_j w_j * F(\sum_{\bar{V}_i \in B_j} \bar{V}_i) = \sum_j \sum_{\bar{V}_i \in B_j} w_j * F(\bar{V}_i)$$

Hence minimizing the value of the objective function is simply a matter of finding the $j$ that minimizes $w_j * F(\bar{V}_i)$ for each feasible $\bar{V}_i$.

**Corollary 1.** *VITA(avg) can be computed exactly in polynomial time.*

*Proof.* Set $\bar{F} = [\frac{1}{d}, \frac{1}{d}, \ldots, \frac{1}{d}]$ where $d$ is the dimension. It is straightforward to see that $\bar{F} \cdot \bar{V} = \frac{\sum_i V_i}{d}$.

Note that many real-world pricing situations are captured by linear F, such as charging separately for the usage of each resource (dimension).

## 3    VITA(min)

### 3.1    Unconstrained, Bounded - Exact

First, we prove two lemmas about the optimal solution which will help us constrain the search space for our exact algorithm.

Without loss of generality assume that the bucket index $j$ is sorted in order of increasing weight $w_j$.

**Lemma 2.** *There exists an optimal solution which uses only the first $b$ buckets, for $b \leq d$. Further, let $\min(j)$ be the dimension with the minimum value in bucket $j$; then, the set $\{\min(j)|1 \leq j \leq b\}$ has all distinct elements.*

*Proof.* It is clear that if in a solution two buckets have the same dimension with the minimum value then the bucket with the larger weight can be emptied into the smaller without increasing the value of the objective function. Thus the set of dimensions with the minimum value must be distinct across buckets and therefore the optimal solution need have at most $d$ buckets. It is also clear that if the optimal solution does not involve a bucket $j$ but does involve a bucket $j' > j$ then all the items in bucket $j'$ can be moved to bucket $j$ without increasing the value of the objective function. Thus the optimal solution may consist only of the first $b$ buckets, for $b \leq d$.

We remind the reader that $\bar{V}_i(k)$ denotes the value in the $k$'th position of the vector $\bar{V}_i$.

**Lemma 3.** *There exists an optimal solution in which item $i$ is placed in that bucket $j$ for which $w_j * V_i(\min(j))$ is minimized, amongst the first $d$ buckets.*

*Proof.* Suppose not. Let item $i$ be placed in bucket $j'$. Now if we move it to bucket $j$ then the value of the objective function is changed by $-w_{j'} * V_i(\min(j')) + w_j * V_i(\min(j))$ which by definition is non-positive. Contradiction, and hence proved.

The above two lemmas give rise to a straightforward search, Algorithm 1.

---

**Algorithm 1.** Exact Algorithm for Unconstrained Bounded VITA(min)

---
1: **for** each permutation $\Pi$ of the first $d$ buckets **do**
2:    **for** each load vector $\bar{V}_i$ **do**
3:       Place load vector $\bar{V}_i$ in that bucket $j$ which minimizes $w_{\Pi(j)} * V_i(\min(\Pi(j)))$
4:    Compute the value of the objective function for this permutation
5: Output the best value over all permutations and the corresponding assignment

---

**Theorem 1.** *Unconstrained, Bounded VITA(min) can be computed exactly in time $O(m * n * d!)$.*

*Proof.* The correctness of Algorithm 1 follows from the prior two lemmas. The running time follows from the fact that the algorithm searches over $d!$ permutations and for each permutation it takes $O(m)$ time to assign each of the $n$ load vectors.

## 3.2 Constrained, Bounded - Strongly NP-Hard

**Theorem 2.** *Constrained, Bounded VITA(min) is strongly NP-hard.*

*Proof.* The proof is by reduction from Bin Packing [18] which is strongly NP-hard. In an instance of Bin Packing we are given $m$ bins of the same (constant) size $S$ and a collection of $n$ items $a_i$ such that $\sum_i a_i = m * S$ and we need to decide if these $n$ items can be packed into the $m$ bins.

Given the instance of Bin Packing we create $m$ buckets and $m+n$ load vectors of dimension 2. $m$ of the load vectors are of the form $[S, 0]$ and the vectors are matched up with the buckets so that each such vector is necessarily assigned to its corresponding bucket. Then for each item $a_i$ there is a load vector $[0, a_i]$ and these vectors are unconstrained and can be assigned to any bucket. All weights are set to 1. Now, it is easy to see that the given instance of Bin Packing is feasible if and only if the value of the objective function of VITA(min) is $m * S$.

## 3.3 Unconstrained, Unbounded - Inapproximable

**Theorem 3.** *Unconstrained, Unbounded VITA(min) is inapproximable unless $P = NP$.*

*Proof.* The proof is by reduction from Set Cover [18].

In Set Cover we are given a collection of $m$ sets over a universe of $n$ elements and a number $C$ and we need to decide whether there exists a subcollection of size $C$ that covers all the elements.

We reduce the given instance of Set Cover to Unconstrained, Unbounded VITA(min) as follows: we let $m$ be the dimension size as well as the number of buckets, one for each set. And, for each element $i$, we have an $m$-dimensional load vector:

$$\bar{V}_i(j) = \begin{cases} 1 & \text{if element } i \in \text{set } j \\ \infty & \text{otherwise} \end{cases}$$

We set the weights of $C$ of the buckets to be 1 and the weights of the remaining buckets to be $\infty$.

It is easy to see that the value of the objective function for Unconstrained, Unbounded VITA(min) is $C$ if and only if there exist $C$ sets covering all the elements, otherwise the value of the objective function is $\infty$. Thus, Unconstrained, Unbounded VITA(min) cannot be approximated to any factor.

## 3.4   Constrained, Unbounded - $O(\log n, \log n)$ Bicriteria

Given that the problem is inapproximable (unless P = NP) we relax our expectations and settle for the next best kind of approximation - a bicriteria approximation, [25] where we relax not just the objective function but also the constraints. In this particular situation we will find a solution that uses at most $O(\log n)$ copies of each bucket while obtaining an assignment whose value is no worse than an $O(\log n)$ factor worse than the optimal solution which uses at most 1 copy of each bucket.

Consider the following LP (Linear Program). Let $y_{jk}$ denote the fraction bucket $j$ gives to dimension $k$, and $x_{ijk}$ denote the weight vector $i$ gives to dimension $k$ of bucket $j$.

$$\min \sum_j w_j \sum_i \sum_k x_{ijk} v_{ik} \quad \textbf{min-LP}$$

$$\text{s.t.} \sum_k y_{jk} = 1 \quad \forall j$$

$$\sum_j y_{jk} = 1 \quad \forall k$$

$$x_{ijk} \le y_{jk} \quad \forall i, j, k$$

$$\sum_j \sum_k x_{ijk} \ge 1 \quad \forall i$$

$$x_{ijk} \ge 0 \quad \forall i, j, k$$

$$y_{jk} \ge 0 \quad \forall j, k$$

**Lemma 4.** *The above LP is a valid relaxation of Constrained, Unbounded VITA(min).*

*Proof.* First we need to verify that this LP is a valid relaxation of the original problem. In other words, every solution of the original problem can be translated to the integer solution of this LP. And every integer solution of this LP is a valid solution of the original problem.

Suppose we have a solution of the original problem. Let $\min(j)$ be the minimum dimension of bucket $j$, and $\sigma(i)$ be the bucket assigned for load vector $i$. The value of the objective function for this solution is $\sum_j w_j \sum_{i:\sigma(i)=j} \bar{V}_i(\min(j))$. Now construct the integer solution of the LP. Let

$$y_{jk} = \begin{cases} 1 & \text{if } k = m(j) \\ 0 & \text{otherwise} \end{cases}$$

and

$$x_{ijk} = \begin{cases} 1 & \text{if } j = \sigma(i), \ k = m(j) \\ 0 & \text{otherwise} \end{cases}$$

Because each bucket only has one minimum dimension, the first constraint is satisfied. And each vector is assigned to one bucket, so the second and third constraints are satisfied also. On the other hand, if we have the integer solution, we can assign $\min(j) = k$ and $\sigma(i) = j$ to have a valid solution of the original problem. So there is a one to one relation between the integer solutions of the LP and the solutions of the original problem. Furthermore, the objective function of the LP is the same as the objective function of the original problem. So the optimal integer LP solution must map to the optimal solution of the original problem, and vice versa.

Let $x_{ijk}^*$ and $y_{jk}^*$ be the optimal solution of the LP. The algorithm is as follows.

---

**Algorithm 2.** Bicriteria Approximation for Constrained Bounded VITA(min)

---
1:
2: **for** Each vector **do**
3:     Order its bucket-dimension pair by $y_{jk}^*$ values. And maximize the corresponding $x_{ijk}^*$ values in order. So there will be only one $x_{ijk}^*$ value that is neither equal to $y_{jk}^*$ nor 0.
4:     **if** This $x_{ijk}^*$ value is greater or equal to $\frac{1}{2}y_{jk}^*$, **then**
5:         round it to $y_{jk}^*$
6:     **else**
7:         round it to 0, and double all the previous non-zero $x_{ijk}^*$ values.
8: **for** $\ln \frac{n}{\varepsilon}$ times **do**
9:     **for** Each dimension $k$ in each bucket $j$ **do**
10:         With probability $y_{jk}^*$ make a copy of bucket $j$ in dimension $k$. And assign all the vectors with $x_{ijk}^* = y_{jk}^*$ to this bucket.

---

**Theorem 4.** *Algorithm 2 is an $O(\log n, \log n)$ bicriteria approximation algorithm for Constrained Bounded VITA(min).*

*Proof.* Notice that, in our algorithm we assume that $x_{ijk}^* = y_{jk}^*$ or 0. This is not hard to achieve. For each item, it will order its favorite bin-dimension pair by $y_{jk}^*$ values. And maximize the corresponding $x_{ijk}^*$ values in order. So there is only one $x_{ijk}^*$ value that is not equal to $y_{jk}^*$ value or 0. If this $x_{ijk}^*$ value is greater or equal to $\frac{1}{2}y_{jk}^*$, we can round it to $y_{jk}^*$. Our new objective value is within twice the LP value. If not, we could round it to 0, and double all the previous non-zero $x_{ijk}^*$ values. Then our value is still within twice the LP value. Even if we don't double the previous $x_{ijk}^*$ values, we still have $\sum_{j,k} x_{ijk}^* \geq 1/2$, which we could use to bound the value output by our algorithm.

The expected value of the solution obtained by the (above randomized) Algorithm 2 is exactly the same as the optimum value of the LP. The expected number of copies of each bucket we make is $\sum_k y_{jk} = 1$. And the probability that vector $i$ is not assigned to one of the buckets is: (where $s = m * d$),

$$\Pi_{j,k}(1 - x_{ijk}^*) \leq \left(1 - \frac{\sum_{j,k} x_{ijk}^*}{s}\right)^s = \left(1 - \frac{1}{s}\right)^s \leq e^{-1}$$

So, if we repeat for $t = \ln \frac{n}{\varepsilon}$ times, then

$$Pr[\text{some vector is not assigned}]$$
$$\leq \sum_i Pr[\text{vector } i \text{ is not assigned}] = \frac{n}{e^t} = \varepsilon$$

The expected value of the solution is $OPT_{LP} \cdot \ln \frac{n}{\varepsilon}$. The expected number of copies of a bucket is $\ln \frac{n}{\varepsilon}$. Thus Algorithm 2 gives a $(\log n, \log n)$-approximation to Constrained Bounded VITA(min).

## 4    VITA(max)

Max - Min and Max are very similar, in that for the lower bound we can use the same log-hardness result since min is 0 and for the upper bound we can set the y variable to be greater than the difference of two dimensions for every pair of dimensions.

### 4.1    Unconstrained, Unbounded - Exact

For example, unconstrained, bounded VITA(max) (see Fig. 2) has an exact (polynomial-time) algorithm because a node above, namely unconstrained, unbounded VITA(max) does; further, this result is obviously tight and hence the square has a dotted background. Squares that do not have a dotted background represent open gaps that present opportunities for further research.

**Theorem 5.** Unconstrained, Unbounded *VITA(max) can be computed exactly in time* $O(m + n)$ *time by placing all items into the bucket with the smallest weight.*

*Proof.* We first show that the bucket with the smallest weight will always be used in the optimal solution. If the bucket with smallest weight is not used in the optimal solution, we can always move all the items in one bucket with non-smallest weight to the bucket with the smallest weight to improve the solution.

Now, we show that if we move all the items in the buckets with non-smallest weight to the bucket with smallest weight, the objective value of this new solution will not increase.

To see this, let the bucket $B_0$ with the smallest weight $w_0$. Let the aggregated vector in $B_0$ be $\bar{V}_0$. Let the bucket $B_i$ with a non-smallest weight $w_i$ in the solution, the aggregated vector in $B_i$ be $\bar{V}_i$.

It is easy to see that $w_0 \cdot \max(\bar{V}_0 + \bar{V}_i) \leq w_0 \cdot (\max(\bar{V}_0) + \max(\bar{V}_1)) \leq w_0 \cdot \max(\bar{V}_0) + w_i \cdot \max(\bar{V}_i)$.

Thus, moving all items from $B_i$ to $B_0$ will not increase the objective value of the current solution.

Moving all items to the smallest weighted buckets is optimal.

## 4.2   Constrained, Bounded - Strongly NP-Hard

**Theorem 6.** Constrained, Bounded *VITA(max) is strongly NP-complete even when the number of dimension equals 2.*

*Proof.* We prove by making reduction from bin packing. For $k$ bins with capacity $c$, we correspondingly assign $k$ buckets. As part of input vectors, we will have $k$ 2-dimensional vectors $(c, 0)$. Each of them are strictly constrained to each bucket. Then for each item $i$ with size $s_i$ in the problem of bin packing, we create a 2-dimensional vector $(0, s_i)$ which can be put into any bucket. We further let each bucket have uniform weight of 1. Then there exists $k$ bins that can hold all the items in the bin packing problem if and only if the objective value of this VITA(max) that equals $kc$ is reachable.

## 4.3   Constrained, Unbounded - $\Theta(\log n)$

**Lemma 5.** Constrained, Unbounded *VITA(max) is strongly NP-complete, and can not be approximated within $O(\log n)$.*

*Proof.* We prove by making reduction from set cover. First we let the number of dimensions of input vector in VITA(max) be the number of elements in the set cover problem. For each element $s_i (i = 1 \sim n)$, we correspondingly let vector $\bar{V}_i$ has value one on dimension $i$, has value zero on all the other dimensions. Thus, there are no two element vectors has one value on the same dimension.

Each subset $S_j$ maps to a bucket $B_j$. If element $s_i \in S_j$, then $\bar{V}_i$ can be placed at bucket $B_j$.

Thus, there exists $k$ subsets that cover all the elements if and only if the objective value of this VITA(max) that equals $k$ is reachable.

**Lemma 6.** Constrained, Unbounded *VITA(max) is $O(\log n)$ approximable.*

*Proof (Proof of Lemma 6).* Consider the following LP. Let $x_{ij}$ be the fraction of item $i$ assigned to bucket $j$.

$$\min \sum_{j=1}^{m} w_j * y_j \text{ max-LP}$$

$$\text{s.t.} \quad y_j \geq \sum_{i=1}^{n} x_{ij} \cdot v_{ik} \quad \forall j, k$$

$$\sum_{j=1}^{m} x_{ij} \geq 1 \quad \forall i$$

It is easy to see that this max-LP is a valid relaxation of *constrained, unbounded* VITA(max). Then we need to repeat rounding $\{x_{ij}\}$ $O(\log n)$ times to make sure that all items are placed to some buckets with high probability. The proof is similar to the part in min-LP.

Directly from Lemmas 5 and 6, we get the following.

**Corollary 2.** Constrained, Unbounded *VITA(max) is $\Theta(\log n)$ approximable.*

# 5   VITA($2^{nd}$ max)

We found VITA($2^{nd}$ max) to be a qualitatively harder problem and thus were forced to consider the restricted version where the weights are uniform and the number of buckets exceeds the (bounded) number of dimensions.

## 5.1   Unweighted, Bounded, Unconstrained - Weakly NP-Hard

**Theorem 7.** *Bounded, Unconstrained VITA($2^{nd}$ max) is weakly NP-hard.*

*Proof.* The proof is by reduction from Partition [18]. In an instance of Partition we are given an array of numbers $a_1, a_2, \ldots, a_n$ such that $\sum_{i=1}^{n} a_i = 2B$, and we are required to decide whether there exist a partition of these numbers into two subsets such that the sum of numbers in each subset is $B$.

Given an instance of Partition we reduce it to an instance of Bounded, Constrained VITA($2^{nd}$ max) as follows: our reduction will use 3 dimensions. For each number $a_i$ we construct the load vector $[0, 0, a_i]$. We add another two vectors, $[L, B, 0]$ and $[B, L, 0]$, where $L >> B$, to the collection of vectors. And, there are two (3-dimensional) buckets with uniform weights which we take to be 1. In an optimal assignment vectors $[L, B, 0]$ and $[B, L, 0]$ will be assigned to different buckets because $L >> B$. Thus, the contribution of each bucket is at least $B$ and the value of the objective function is always at least $2B$. Now, from our construction, it is easy to see that if the given instance of Partition has a partition into two subsets with equal sums then the value of the objective function (of the instance) of VITA($2^{nd}$ max) (to which it is reduced) is $2B$. And if there is no equal sum partition into two subsets, then one of the buckets necessarily has a $2^{nd}$ max dimension value greater than $B$, which means that the objective value has to be larger than $2B$.

## 5.2   Unweighted, Constrained, with Number of Buckets Exceeding Number of Dimensions - $O(\log n)$ Approximation

Consider the following LP. Let $x_{ij}$ be the fraction of vector $i$ assigned to bucket $j$.

$$\min \sum_{j=1}^{m} y_j \quad 2^{nd}\text{max-LP}$$

$$\text{s.t.} \ \ y_j \geq \sum_{i=1}^{n} x_{ij} \cdot v_{ik} \ \ \forall j, k \ (j \neq k)$$

$$\sum_{j=1}^{m} x_{ij} \geq 1 \qquad \forall i$$

**Lemma 7.** *The above LP $2^{nd}$ max-LP is a valid relaxation of constrained VITA($2^{nd}$ max) where the number of buckets exceeds the number of dimensions.*

*Proof.* First we need to verify that $y_j$ really represents the $2^{nd}$-maximum dimension in the LP solution. From the first LP constraint, we know $y_j$ is either the maximum dimension or the $2^{nd}$-maximum dimension. The following proof shows that based on the current LP optimum we could come up with a new LP optimum solution in which $y_j$ is the $2^{nd}$-maximum dimension of bin $j$. For each bin $j$ with $y_j$ as maximum dimension, there are only 2 cases, as follows.

*Case 1: the item, with $y_j$'s corresponding dimension as "free" dimension, has its "free" dimension as maximum.* In bin $j$ the "free" dimension is $j^{th}$ dimension. Assume $y_j$ represents the value in dimension $d_j$ of bin $j$, then we can find the bin in which dimension $d_j$ is the maximum ("free" dimension). Merge these two bins together and set $d_j$ as the "free" dimension of this bin. In the new solution, the cost won't be more than the previous optimal solution, which means this is also an optimal solution.

*Case 2: the item, with $y_j$'s corresponding dimension as "free" dimension, doesn't have its "free" dimension as maximum.* Let bin $j$ have "free" dimension $j$. $y_j$ represents the value of dimension $d_j$ of bin $j$ and it is the maximum dimension. Bin $k$ has $d_j$ as "free" dimension. And $y_k$ is the maximum dimension of bin $k$. Then swap these two bins. The cost of new bin $k$ is less than $y_j$ and the cost of new bin $j$ is at most equal to $y_k$. So the cost of new solution is better than original optimal solution. This is a contradiction, which means this case couldn't happen.

    To sum up, given an optimal solution of the LP, we can come up a new optimal solution in which each $y_j$ represents the $2^{nd}$-maximum dimension of bin $j$.

**Lemma 8.** *Unweighted, Constrained, VITA($2^{nd}$ max) with number of buckets exceeding number of dimensions can be approximated to factor $O(\log n)$.*

*Proof.* As with the algorithm and proof for min-LP, we need to repeat rounding $\{x_{ij}\}$ $O(\log n)$ times to make sure that all vectors are placed in some bucket with high probability.

## 6    Experiments

We implemented *LP-Approx* and the three heuristics in Python, using Python 2.7.5. We use SageMath [31] and GLPK [8] as our Linear Programming Solver. We conducted our experiments on a single core of a 4-core Intel i7-3770 clocked at 3.4 GHz (0.25 MB L2 cache per core, and 2 MB L3 cache per core), with 16 GiB of DDR3-1600 RAM.

Nutanix is a vendor of hyper-converged infrastructure appliances. For this paper we used a dataset obtained from an in-house cluster they maintain for testing and validation purposes. The cluster runs real customer workloads. The data was logged using the Prism system of Nutanix and then filtered, anonymized and aggregated before being handed to us. The dataset we received comprised of measurements logged every 5 min of CPU, memory and storage used by 643 different services running continuously for the entire calendar month of August 2017. The data consisted of $643 \times 8928$ rows of 6 columns - timestamp, serviceID, CPU-usage, memory-usage, storage-utilization and bandwidth-usage.

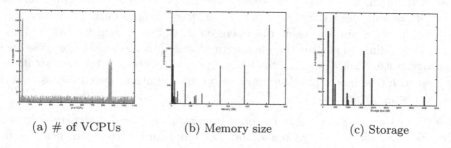

(a) # of VCPUs          (b) Memory size          (c) Storage

**Fig. 3.** Histograms of requested resources

---

**Algorithm 3.** Heuristic 1 - Conservative

---

1: **for** each vector **do**
2:    Assign the vector $V_i$ to that bucket $j$ which minimizes $w \cdot F(V_i)$.

---

The goal of our experiments was to compare the LP-based approximation algorithms to 3 natural polynomial-time heuristics - *Conservative*, *Greedy* and *Local-Search* - on each of the 4 problems - VITA(max), VITA(min), VITA(max − min) and VITA($2^{nd}$ max). We briefly describe the 3 heuristics:

– *Conservative* This heuristic assigns each vector in isolation, i.e. it assigns each vector $\bar{V}_i$ to that bucket $j$ which minimizes $w_j \cdot \tilde{F}(\bar{V}_i)$.
– *Greedy* The heuristic detailed in Algorithm 4 selects the vectors one by one in a random order and assigns to the bucket that minimizes the increase in the objective value.
– *Local-Search* Local search based vector placement in Algorithm 5 starts from a random feasible placement and repeatedly swaps vectors between two buckets to decrease the objective value. Since the size of the potential search space is exponential in $n$, the number of vectors, we restrict the heuristic to run the swapping step for a linear number of times.

It is easy to see that all the 3 heuristics can be arbitrarily bad $(\Omega(n))$ in terms of quality of approximation. However, we are interested in comparing their behavior on practical workloads vis a vis each other as well as the corresponding LP-based approximation algorithm. We run each of the 4 schemes (3 heuristics and 1 approximation algorithm) on samples of $n$ vectors drawn from the dataset. Each sample is drawn uniformly from the entire dataset $n$ runs from 10 to 100 in steps of 10. Given a sample we simulate each scheme on the sample to obtain a measure of the solution-quality and run-time[5]. For a given $n$ we run as many samples as are needed to minimize the sample variance of the statistic (solution-quality or run-time) to below 1% of the sample mean. For VITA(max) we utilize the 3 dimensions - CPU, memory and storage - after a suitable normalization, and averaged over the entire month, i.e. we sample from 643 rows. For VITA(min) we aggregate CPU usage on an hourly basis (from the 5 min measurements which reduces the dataset from 8928 to 744 rows per service). For VITA(max − min) we aggregate bandwidth usage on an hourly basis per service. For VITA($2^{nd}$ max) we use the bandwidth usage on a 5 min basis for each service. Based on our experiments we collected measurements on the two main considerations - (1) solution quality and, (2) running time, for each of VITA(max), VITA(min), VITA(max − min) and VITA($2^{nd}$ max). In Figs. 2, 3, 4 and 5 we use VITA(f) in place *LP-Approx* to emphasize the specific function f under consideration.

---

**Algorithm 4.** Heuristic 2 - Greedy

---

1: Shuffle the order of vectors;
2: **for** each vector **do**
3:     Assign the vector to that bucket such that the current objective value is raised the least;

---

---

**Algorithm 5.** Heuristic 3 - Local-Search

---

1: **for** each vector **do**
2:     Randomly assign it to a feasible bucket by affinity constraint;
3: **for** 1 to $poly(n)$ steps **do**
4:     **for** every two buckets **do**
5:         Swap any pair of two vectors if the swap will reduce the objective value;

---

## 6.1  Solution Quality

From Fig. 4a, c and d, it can be seen that the linear programming based approximation outperforms the heuristics for VITA(max), VITA(max − min) and VITA($2^{nd}$ max) by a factor of about 1.5. Unfortunately, the out-performance

---

[5] We do not implement these schemes in the Nutanix system and then measure their performance as that would be expensive in terms of development time and would produce little additional clarity over the simulation based approach.

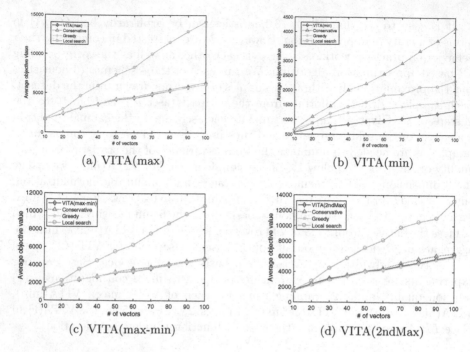

(a) VITA(max)    (b) VITA(min)

(c) VITA(max-min)    (d) VITA(2ndMax)

**Fig. 4.** Quality of approximation of VITA(max, min, max-min, 2ndMax) vs {*Greedy, Conservative, Local-Search*}

does not stand out visually because of the compression in the scale of the graph caused by the very poor performance of *Local-Search*. *Local-Search* performs particularly poorly in these 3 cases due to its dependence on the starting configuration.

For *minimizing the maintenance down time* in Fig. 4b, VITA(min) performs better than any of *Greedy*, *Local-Search* and *Conservative*. This is because VITA(min)'s bicriteria approximation scheme allows for the use of additional buckets, see Fig. 6. However, when the same number of extra buckets are given to all heuristics, we see that *Greedy* performs best.

## 6.2   Running Time

Here we focus only on VITA and *Greedy* for two reasons: (1) Previous experiment results on solution quality show that VITA and *Greedy* are the two approaches of interest (2) *Local-Search* has much higher run time complexity than others. Fig. 5a–d show that *Greedy*, basically linear-time, is superior to the LP based approximation algorithms (Fig. 7).

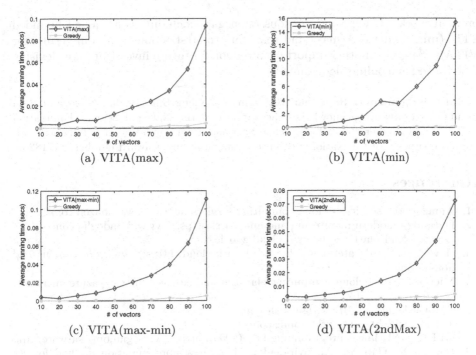

(a) VITA(max)    (b) VITA(min)

(c) VITA(max-min)    (d) VITA(2ndMax)

**Fig. 5.** Running time of VITA(max, min, max-min, 2ndMax) vs *Greedy*

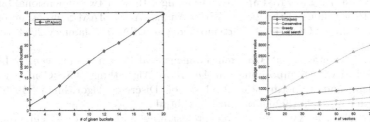

**Fig. 6.** # of used buckets vs # of given buckets for VITA(min)

**Fig. 7.** Solution quality with same number of additional buckets given to heuristics

## 7    Conclusion and Future Work

We have proposed a new and general framework VITA that captures several naturally occurring problems in the context of hybrid clouds. We presented novel hardness results and approximation algorithms (using clever LP rounding). We conducted a detailed experimental evaluation comparing our approximation algorithm to several natural heuristics.

On the experimental side it would be interesting to characterize natural work-loads and develop heuristics with provable (average-case) guarantees. Our theoretical work has left some obvious open gaps including constrained bounded VITA(min) and VITA(max) and removing the restrictions from our results for VITA($2^{nd}$ max). Another important direction for future investigation is devising distributed and online algorithms.

**Acknowledgements.** Rajmohan Rajaraman, Zhifeng Sun, Bochao Shen and Ravi Sundaram gratefully acknowledge the support of the National Science Foundation under award number #1535929. Bochao Shen and Ravi Sundaram gratefully acknowledge the support of the National Science Foundation under award number #1718286.

# References

1. Amazon web services - cloud computing services. https://aws.amazon.com/
2. Datadog - modern monitoring and analytics. https://www.datadoghq.com/
3. Google cloud platform. https://cloud.google.com/
4. Hewlett packard enterprise - hybrid it with cloud. https://www.hpe.com/us/en/home.html
5. Microsoft azure cloud computing platform and services. https://azure.microsoft.com/en-us/
6. Rightscale. https://www.rightscale.com/
7. Turbonomic. https://turbonomic.com/
8. GLPK (GNU linear programming kit) (2006). http://www.gnu.org/software/glpk
9. Bansal, N., Caprara, A., Sviridenko, M.: A new approximation method for set covering problems, with applications to multidimensional bin packing. SIAM J. Comput. **39**(4), 1256–1278 (2009)
10. Bansal, N., Khan, A.: Improved approximation algorithm for two-dimensional bin packing. In: Proceedings of the Twenty-Fifth Annual ACM-SIAM Symposium on Discrete Algorithms, SODA 2014, Portland, Oregon, USA, 5–7 January 2014, pp. 13–25 (2014)
11. Chekuri, C., Khanna, S.: On multi-dimensional packing problems. In: SODA: ACM-SIAM Symposium on Discrete Algorithms (A Conference on Theoretical and Experimental Analysis of Discrete Algorithms) (1999). citeseer.ist.psu.edu/chekuri99multidimensional.html
12. Chen, G., et al.: Energy-aware server provisioning and load dispatching for connection-intensive internet services. In: Proceedings of the 5th USENIX Symposium on Networked Systems Design and Implementation, NSDI 2008, pp. 337–350 (2008)
13. Christensen, H.I., Khan, A., Pokutta, S., Tetali, P.: Multidimensional bin packing and other related problems: a survey. https://people.math.gatech.edu/~tetali/PUBLIS/CKPT.pdf
14. Dimitropoulos, X., Hurley, P., Kind, A., Stoecklin, M.P.: On the 95-percentile billing method. In: Moon, S.B., Teixeira, R., Uhlig, S. (eds.) PAM 2009. LNCS, vol. 5448, pp. 207–216. Springer, Heidelberg (2009). https://doi.org/10.1007/978-3-642-00975-4_21
15. Dupont, C., Hermenier, F., Schulze, T., Basmadjian, R., Somov, A., Giuliani, G.: Plug4Green: a flexible energy-aware VM manager to fit data centre particularities. Ad Hoc Netw. **25**, 505–519 (2015)
16. Dupont, C., Schulze, T., Giuliani, G., Somov, A., Hermenier, F.: An energy aware framework for virtual machine placement in cloud federated data centres. In: e-Energy, p. 4. ACM (2012)

17. Frieze, A., Clarke, M.: Approximation algorithms for the m-dimensional 0-1 knapsack problem: worst-case and probabilistic analyses. Eur. J. Oper. Res. **15**(1), 100–109 (1984)
18. Garey, M.R., Johnson, D.S.: Computers and Intractability: A Guide to the Theory of NP-Completeness. W. H. Freeman & Co., New York (1979)
19. Ghodsi, A., Zaharia, M., Hindman, B., Konwinski, A., Shenker, S., Stoica, I.: Dominant resource fairness: fair allocation of multiple resource types. In: Proceedings of the 8th USENIX Symposium on Networked Systems Design and Implementation, NSDI 2011, Boston, MA, USA, 30 March–1 April 2011 (2011)
20. Herbst, N.R., Kounev, S., Reussner, R.: Elasticity in cloud computing: what it is, and what it is not. In: Proceedings of the 10th International Conference on Autonomic Computing (ICAC 2013), pp. 23–27. USENIX, San Jose (2013). https://www.usenix.org/conference/icac13/technical-sessions/presentation/herbst
21. Hermenier, F., Lawall, J.L., Muller, G.: BtrPlace: a flexible consolidation manager for highly available applications. IEEE Trans. Dependable Sec. Comput. **10**(5), 273–286 (2013)
22. Hochbaum, D., Shmoys, D.: Using dual approximation algorithms for scheduling problems: theoretical and practical results. J. ACM **34**, 144–162 (1987). citeseer.ist.psu.edu/470961.html
23. van den Hooff, J., Lazar, D., Zaharia, M., Zeldovich, N.: Vuvuzela: scalable private messaging resistant to traffic analysis. In: Proceedings of the 25th Symposium on Operating Systems Principles, SOSP 2015, Monterey, CA, USA, 4–7 October 2015, pp. 137–152 (2015)
24. Jansen, R., Bauer, K.S., Hopper, N., Dingledine, R.: Methodically modeling the tor network. In: 5th Workshop on Cyber Security Experimentation and Test, CSET 2012, Bellevue, WA, USA, 6 August 2012 (2012)
25. Marathe, M.V., Ravi, R., Sundaram, R., Ravi, S.S., Rosenkrantz, D.J., Hunt III, H.B.: Bicriteria network design problems. J. Algorithms **28**(1), 142–171 (1998)
26. Mathewson, N., Dingledine, R.: Practical traffic analysis: extending and resisting statistical disclosure. In: Martin, D., Serjantov, A. (eds.) PET 2004. LNCS, vol. 3424, pp. 17–34. Springer, Heidelberg (2005). https://doi.org/10.1007/11423409_2
27. Raghavan, P., Thompson, C.D.: Randomized rounding: a technique for provably good algorithms and algorithmic proofs. Combinatorica **7**(4), 365–374 (1987). https://doi.org/10.1007/BF02579324
28. Reddyvari Raja, V., Dhamdhere, A., Scicchitano, A., Shakkottai, S., Claffy, Kc., Leinen, S.: Volume-based transit pricing: is 95 the right percentile? In: Faloutsos, M., Kuzmanovic, A. (eds.) PAM 2014. LNCS, vol. 8362, pp. 77–87. Springer, Cham (2014). https://doi.org/10.1007/978-3-319-04918-2_8
29. Raja, V.R., Shakkottai, S., Dhamdhere, A., Claffy, Kc.: Fair, flexible and feasible ISP billing. SIGMETRICS Perform. Eval. Rev. **42**(3), 25–28 (2014)
30. Stillwell, M., Vivien, F., Casanova, H.: Virtual machine resource allocation for service hosting on heterogeneous distributed platforms. In: 26th IEEE International Parallel and Distributed Processing Symposium, IPDPS 2012, Shanghai, China, 21–25 May 2012, pp. 786–797 (2012)
31. The Sage Developers: SageMath, the Sage Mathematics Software System (Version 8.1) (2017). http://www.sagemath.org
32. de la Vega, W.F., Lueker, G.S.: Bin packing can be solved within $1+\epsilon$ in linear time. Combinatorica **1**, 349–355 (1981)

# A Peer-to-Peer Based Cloud Storage Supporting Orthogonal Range Queries of Arbitrary Dimension

Markus Benter[1]($\boxtimes$), Till Knollmann[2]($\boxtimes$), Friedhelm Meyer auf der Heide[2]($\boxtimes$),
Alexander Setzer[3]($\boxtimes$), and Jannik Sundermeier[2]($\boxtimes$)

[1] Jobware GmbH, Technologiepark 32, 33100 Paderborn, Germany
`m.benter@jobware.de`
[2] Computer Science Department and Heinz Nixdorf Institute,
Paderborn University, Fürstenallee 11, 33102 Paderborn, Germany
`{tillk,fmadh,janniksu}@mail.uni-paderborn.de`
[3] Computer Science Department, Paderborn University, Paderborn, Germany
`asetzer@mail.uni-paderborn.de`
`https://www.hni.uni-paderborn.de/alg/`, `https://cs.uni-paderborn.de/ti/`

**Abstract.** We present a peer-to-peer network that supports the efficient processing of orthogonal range queries $R = \times_{i=1}^{d}[a_i, b_i]$ in a $d$-dimensional point space. The network is the same for each dimension, namely a distance halving network like the one introduced by Naor and Wieder (ACM TALG'07). We show how to execute such range queries using $\mathcal{O}\left(2^{d'} d \log m + d |R|\right)$ hops (and the same number of messages) in total. Here $[m]^d$ is the ground set, $|R|$ is the size and $d'$ the dimension of the queried range. Furthermore, if the peers form a distributed network, the query can be answered in $\mathcal{O}\left(d \log m + d \sum_{i=1}^{d}(b_i - a_i + 1)\right)$ communication rounds. Our algorithms are based on a mapping of the Hilbert Curve through $[m]^d$ to the peers.

**Keywords:** Distributed storage · Multi-dimensional range queries ·
Peer-to-Peer · Hilbert Curve

## 1 Introduction

Consider a scenario in which the content of a music sharing platform is distributed among the participants of a peer-to-peer network (P2P network). Classical P2P networks only consider search queries for a specific attribute of the data contained in the network. For a music sharing platform, however, a crucial requirement is to allow more complex search queries. Users typically want

This work was partially supported by the German Research Foundation (DFG) within the Collaborative Research Center 'On-The-Fly Computing' (SFB 901).

Y. Disser and V. S. Verykios (Eds.): ALGOCLOUD 2018, LNCS 11409, pp. 46–58, 2019.
https://doi.org/10.1007/978-3-030-19759-9_4

to filter the data by different attributes and, most important, filter by *multiple* attributes. A typical request would be to find all jazz, funk & soul songs published in the last two years. Further queries could involve the language, the audio quality of the tracks or the duration. In the last years, several publications investigate the design of P2P networks allowing for multi-dimensional range queries. For more details, we refer the reader to Sect. 1.1. Most of these solutions, however, involve very complex network designs and the design depends heavily on the dimensionality of the data stored in the network. In this work, we aim at a more lightweight solution which uses the same network structure independent of the dimensionality of the data. Additionally, we are interested in an analysis of the message complexity and the delay for answering multi-dimensional range queries. More formally, the topic of this work can be described as follows:

Consider a distributed system storing a set of data items associated with keys. The keys are given by multiple attributes of the data items such that they can be seen as points in a $d$-dimensional mesh. We consider the $d$-dimensional Range Query Problem defined as follows.

**Definition 1 ($d$-dimensional Range Query Problem)** *Let $M(m, d)$ be a $d$-dimensional mesh with side-length $m$ where $m = 2^k$ for a $k \in \mathbb{N}$. Let $R = \times_{i=1}^{d}[a_i, b_i]$ be an orthogonal range (or range for short) in $M(m, d)$. The $d$-dimensional Range Query Problem is the task of reporting all points in $R$. The dimension of a range, $d'$, is the number of all dimensions $i$ with $a_i \neq b_i$.*

Our goal is to distribute the $m^d$ points of $M(m, d)$ over $m^d$ peers of a P2P network such that range queries can be answered using few hops and few communication rounds. In a communication round, each peer can receive and send a set of messages. Messages sent in communication round $i$ are all received in round $i + 1$. The solution can be scaled to fewer peers in a simple way, as mentioned in the final section. The main focus of this work is to realize range queries of arbitrary dimension on one fixed network, rather than using a specific one for each dimension. In addition, we aim at a topology with constant degree.

Let $|R| = \prod_{i=1}^{d}(b_i - a_i + 1)$ be the size of a range $R$ in our mesh. A trivial topology for the $d$-dimensional Range Query Problem would be the $d$-dimensional grid. This topology allows to answer a query for $R$ in $\mathcal{O}(c + |R|)$ hops where $c$ is the distance from the entry peer to the closest point in $R$. However, the grid topology depends on the dimension of the keys. Another trivial solution when assuming a constant degree network would be to query each point in $R$ separately. While the network is independent of the mesh, the total number of hops can only be bounded to $\mathcal{O}(|R| \cdot d \cdot \log m)$ due to each constant degree network with $m^d$ nodes having a diameter of $\Omega(d \cdot \log m)$. The main question we deal with in this paper is how close we can come to the bound of $\mathcal{O}(c + |R|)$ while still having a constant degree network topology.

The solution we present in this work uses a Hilbert Curve that maps the $m^d$ points of $M(m, d)$ to a one-dimensional line. We assume a very dense set of keys, i.e., all points in $M(m, d)$ refer to existing keys stored in the system. As we consider $m^d$ points, we assume a bijective mapping from $[m]^d$ to the peers such that the Hilbert Curve defines an ordering of the peers. In Sect. 2, we present the topology of our P2P network, namely the Distance Halving Graph, the mapping of the nodes of $M(m, d)$ on it, and properties of this mapping. We present an algorithm for answering a range query $R$ in Sect. 3 and prove that the total number of hops needed to answer a range query $R$ is $\mathcal{O}\left(2^{d'} d \log m + d |R|\right)$. Moreover, we show that the algorithm can be parallelized and then answers a range query in $\mathcal{O}\left(d \log m + d \sum_{i=1}^{d}(b_i - a_i + 1)\right)$ communication rounds.

## 1.1  Related Work

Recently, the design of P2P networks for the purpose of answering multi-dimensional range queries has attracted much attention in the research community. For one-dimensional range queries, CAN, P-Grid, Baton, Armada, Saturn, and PHT have been proposed [3, 6, 9, 11, 14, 15]. Throughout this section, let $n$ be the number of peers in the network. Among these, Baton and Armada can answer one-dimensional range queries in time $\mathcal{O}(\log n + |R|)$, where $|R|$ denotes the number of data items in the range $R$. This is asymptotically optimal [11]. Baton, however, has a logarithmic degree at each node, whereas Armada requires only a constant degree.

For multi-dimensional ranges, there is no approach known which comes close to the optimal bound while having only a constant degree. In the literature, most work focuses on an experimental evaluation of P2P networks allowing for multi-dimensional range queries. In this work, we aim at a rigorous analysis of the message complexity and the communication rounds needed for answering a range.

P2P networks designed for multi-dimensional range queries can be subdivided into classes based on their approaches of mapping the data space and connect the peers to each other. MURK uses a k-d tree to partition the data space such that each node of the network is responsible for exactly one hypercuboid of the data space [7]. Moreover, MURK supports efficient routing by adding random links or a space-filling Skip Graph. Both approaches, however, require $\mathcal{O}(\log n)$ references stored at each node such that the resulting degree of the network is $\mathcal{O}(\log n)$.

Other approaches using Skip Graphs, are SkipIndex [20], ZNet [18] and SCRAP [7]. The difference in these approaches is the way of partitioning the data (for instance a space-filling $Z$-curve in Znet) and the way of choosing additional overlay edges for efficient routing. As all these approaches are based on Skip Graphs, the degree of each network is $\mathcal{O}(\log n)$.

MatchTree [10] uses an interesting approach which builds an individual tree for each query. The underlying P2P network is a variant of Kleinberg's small world network [12]. In these networks, each node has $\mathcal{O}(\log n)$ shortcut neighbors chosen randomly. Since the tree for each query is built at the time of processing the query, each node has to store a lot of structural information. In addition, lots of nodes are involved in answering a query.

Another approach working with trees is DRAGON [5]. In DRAGON, the identifiers of peers are distributed in $[0, 1)$ and an aggregation tree is built upon the interval $[0, 1)$. The structure of the aggregation tree depends on the space filling $Z$-curve such that each node is responsible for a certain region of the curve. To ensure efficient query processing, each node stores a reference to each level of the aggregation tree in its local routing table which also results in a degree of $\mathcal{O}(\log n)$.

SWORD [2] and MAAN [4] are architectures which use multiple hash functions, one for each dimension of the data, and map each of these hash functions onto the same network. MAAN uses a locality preserving hash function for each of the $d$ dimensions per resource. A triple of $<dimension, value, resource\text{-}info>$ is stored at each node that stores a dimension of a resource due to a hash function. The authors present two routing algorithms for resolving ranges. The first one does a range query for each dimension and returns the cut of all results. While the cost for routing to the peer at the beginning of each range is in a total of $\mathcal{O}(d \log n)$ hops, the cost for gathering all values in a range concerning only one dimension results in $\mathcal{O}(m^{d-1} \sum_{i=1}^{d}(b_i - a_i + 1))$ hops in our scenario. Further, due to the splitting of the dimension set of each resource, an additional total memory of $\mathcal{O}(d m^d)$ is needed. The second routing algorithm improves the routing time at the cost of memory requirement. Each resource is stored $d$ times in total. The routing then only processes a range query along one specific dimension and removes all false-positives. This induces a routing time in $\mathcal{O}(\log n + \min_i\{(b_i - a_i + 1)\})$ while the memory demand increases by a factor of $d$. In contrast to MAAN, we focus on querying exactly those points which are in the range, i.e., we do not have an (intermediate) over-approximation of the results. Further, our solution does not increase the memory requirement for storing keys.

LORM [17] also uses a distributed hash table. In LORM, the peers are organized in clusters and each cluster of peers is responsible for a single dimension. To answer queries, the original query is split into subqueries for each dimension which are then answered by the affected clusters. The worst case bounds for identifying nodes which are involved in a range query of LORM are similar to the bound mentioned for MAAN. To achieve the bounds, LORM also relies on a logarithmic network degree.

The approach closest to ours, called Squid, combines the Chord-architecture [19] with the space-filling Hilbert Curve [16]. The multi-dimensional space is mapped onto the one-dimensional space by using a Hilbert Curve. Queries are answered by recursively refining a query and map subqueries to peers of the network. Although the authors prove that only a small number of nodes is contacted while query processing, it is not clear how many messages are sent in the process and which delay is caused by answering a query.

## 2   Notations

The ground set of elements to be queried in the $d$-dimensional range query problem is the node set $[m]^d$ of the $d$-dimensional mesh $M(m, d)$ with side-length $m$. For ease of description, we assume $m = 2^k$ for $k \in \mathbb{N}_0$. Fix some $i$, $0 \le i \le k$, and let $m' = \frac{m}{2^i}$. $M(m,d)$ can be divided into $2^{di}$ meshes with side-lengths $m'$. These meshes are denoted as *canonical submeshes* with side-lengths $m'$ (c.f. Fig. 1).

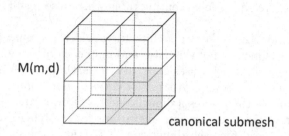

**Fig. 1.** The subdivision of a 3-dimensional mesh into $2^3 = 8$ canonical submeshes with side-lengths $m' = \frac{m}{2}$.

## 2.1   Embedding a Hilbert Curve in a Mesh

The $M(m,d)$-*Hilbert Curve* is a curve that connects all points of $M(m,d)$. It is a discrete version of the space-filling Hilbert Curve [8]. The $M(2,2)$-Hilbert Curve is defined as in the left picture of Fig. 2. The $M(m,2)$-Hilbert Curve can be subdivided into four rotated $M(\frac{m}{2},2)$-Hilbert Curves. The upper left and upper right $M(\frac{m}{2},2)$-Hilbert Curves are not rotated. The lower left one is rotated by $90°$ clockwise and the lower right one is rotated by $90°$ counterclockwise.

The $M(m,d)$-Hilbert Curve is an extension of the $M(m,2)$-Hilbert Curve. For $d$ dimensions, a $M(m,d)$-Hilbert Curve can be subdivided into $2^d$ $M(\frac{m}{2},d)$-Hilbert Curves. For the formal definition, we refer the reader to [1]. The definition of the $M(m,d)$-Hilbert Curve ensures that all points of a canonical submesh are connected by a Hilbert Curve of smaller size. Further, we define an ordering of all points in $M(m,d)$ by the order in which they are visited via the $M(m,d)$-Hilbert Curve starting at the origin. Then the *Hilbert Coordinate* $p(v)$ of $v \in M(m,d)$ is the position of $v$ in this ordering.

## 2.2   Network Properties

The P2P network we consider is the Distance Halving Graph as defined in [13]. We use the version with perfect smoothness. In this case, we can simply assume that the ids of the $n$ peers $\{w_0, \ldots, w_{n-1}\}$ are $\{0, \ldots, n-1\}$ instead of $\{\frac{0}{n}, \ldots, \frac{n-1}{n}\}$. The peer $w_i$ has undirected edges to its direct neighbors $w_{i-1}$ and $w_{i+1}$ (for $i > 0$ and $i > n-1$, resp.). In addition, it is connected to the peers with ids $\lfloor \frac{i}{2} \rfloor$ and $\lfloor \frac{i}{2} \rfloor + \lfloor \frac{n}{2} \rfloor$ (if $\lfloor \frac{i}{2} \rfloor + \lfloor \frac{n}{2} \rfloor \le n - 1$). We consider each edge to be undirected. It has been shown in [13] that the Distance Halving Graph with perfect smoothness has a constant degree. Lemma 1 is a direct consequence of the above construction.

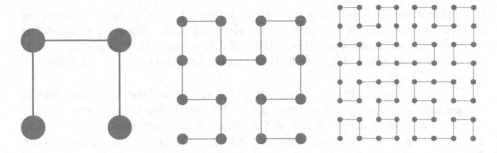

**Fig. 2.** M(m,d)-Hilbert Curve of dimension $d = 2$ and $m = 2, 4$ and 8.

**Lemma 1.** *The Distance Halving Graph supports a routing between $w_i$ and $w_j$ using at most $3 \cdot \log(|j - i| + 1)$ hops.*

*Proof.* Due to the definition of the Distance Halving Graph, $w_i$ and $w_j$ have edges to nodes $w_{i'}$ and $w_{j'}$ such that $|i' - j'| \leq \lfloor \frac{i}{2} \rfloor - \lfloor \frac{j}{2} \rfloor \leq \frac{1}{2}|i - j| + 1$. Applying an appropriate move of $w_i$ to a direct neighbor reduces the distance to at most $\leq \frac{1}{2}|i - j|$. Iterating these three hops $\lfloor \log(|i - j| + 1) \rfloor$ times yields the lemma. (Note: We have only used one kind of the edges to non-direct neighbors. Combining both kinds of hops can be used to reduce the congestion of the routing, see [13].) $\qquad\square$

Now assume that $n$, the size of the P2P network, equals $m^d$, the size of our ground set. We define the one-to-one mapping from $M(m, d)$ to the peers by the ordering induced by the Hilbert Curve: Node $v$ from $M(m, d)$ is mapped to the peer with id $p(v)$. As the nodes in a canonical submesh of $M(m, d)$ are ordered consecutively in the ordering induced by the Hilbert Curve, by Lemma 1 we can conclude:

**Observation 1.** *In a canonical submesh with side-length $m'$, routing between any two points costs at most $\mathcal{O}(d \cdot \log(m') + 1)$ hops.*

## 3  Answering Range Queries

Whenever we talk about a range, we mean an orthogonal range defined as follows. An orthogonal range is given by $R = \times_{i=1}^{d}[a_i, b_i]$. We call the set of dimensions $i$ in which $a_i < b_i$ the *extensions* of $R$, denoted by $D$. The number of extensions is the dimension of $R$ denoted by $|D| = d'$.

A point $x = (x_1, \ldots, x_d)$ is a *corner point* of $R$ if, for each $i$, $x_i = a_i$ or $x_i = b_i$ holds. Clearly, a range $R$ with dimension $d'$ has $2^{d'}$ corner points. Especially, a one-dimensional range, a *line $L$*, has two corner points. We say a line $L$ *crosses a range $R$* if $L$ only consists of points in $R$ and has maximal length.

Let $R$ be a $d'$-dimensional range. $R$ is contained in a $d'$-dimensional Mesh $M_R$ whose points are all points $(x_1, \ldots, x_d)$ of $M(m, d)$ with $x_i = a_i$ for all $i \notin D$.

We define canonical submeshes of $M_R$ as intersections of canonical submeshes of $M(m, d)$ with $M_R$. From now on, we always consider $M_R$, i.e., if not stated otherwise, a canonical submesh is of $M_R$ and dimensions which are not mentioned are assumed to match $M_R$. We further assume that dimensions 1 to $d'$ are the dimensions in $D$.

Consider a line $L$ with extension $i$ such that $a_i < b_i$ and the smallest canonical submesh $S = \times_{j \in D}[x_j, y_j]$ surrounding $L$. We call $L$ *touching*, if either $a = x_i$ or $b = y_i$ is a point at the border of $S$ in dimension $i$. Note that, for a touching line $L$, the smallest enclosing canonical submesh has side-length at most $2|L|$. Consider a range $R$ and its set of extensions $D$. $R$ is *touching*, if every line $L$ with an extension $i \in D$ that crosses $R$ is touching (Figs. 3 and 4).

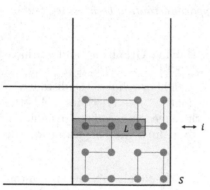

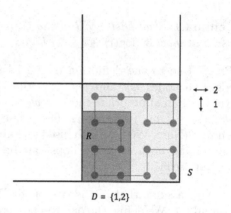

**Fig. 3.** A touching line query. $L$ touches the canonical submesh $S$ in dimension $i$.

**Fig. 4.** A touching range query $R$ in the canonical submesh $S$. The extension set of $R$ is $D = \{1, 2\}$.

A major problem we have to deal with when answering range queries is the following: There are small ranges, whose smallest enclosing canonical submesh is large, may even be the entire mesh $M(m, d)$. An extreme range consists of the two nodes $(0, ..., 0, \frac{m}{2} - 1)$ and $(0, ..., 0, \frac{m}{2})$. See Fig. 5 for example. If we try to report such ranges following the Hilbert Curve, we have to follow many long routing paths between elements from the range, resulting in a large hop count. To deal with this problem, we will partition a range in subranges, each of which can be reported using only short such paths.

This partition is defined as follows. The *canonical box around $R$* is:

$$C(R) = \underset{i \in D}{\times} C_i, \text{ with } C_i = [c_i\, 2^j, (c_i + 1)\, 2^j - 1]$$

where $c_i \in \mathbb{N}_0$ and $j \geq 0$ minimal such that $c_i\, 2^j \leq a_i \leq b_i \leq (c_i + 1)\, 2^j - 1$.

The center $z$ of $C(R)$ is

$$z = (z_1, \ldots, z'_d) \text{ with } z_i = \frac{c_i\, 2^j + (c_i + 1)\, 2^j - 1}{2}.$$

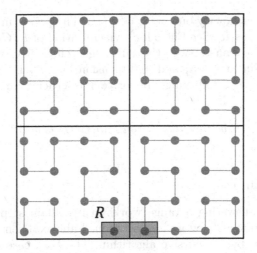

**Fig. 5.** Example of a small range $R$ having a large smallest enclosing canonical submesh. The points of $R$ are direct neighbors in $M(m,d)$ but far away from each other on the Hilbert Curve.

Consider the $2^{d'}$ orthants $O_1, \ldots, O_{2^{d'}}$ on $M_R$ centered around $z$. For example, one of them is given by the points $\{x \in M_R \mid x_i > z_i \forall i \in D\}$. Now consider the subranges $R_i = R \cap O_i$. The following observation is crucial for our algorithm:

**Observation 2.** *Every subrange $R_i$ is touching.*

Now consider the set $Z(R)$ of points of $M_R$ centered around $z$:

$$Z(R) = \{p \mid p_i = z_i \pm (1/2) \text{ for } i \in D\}$$

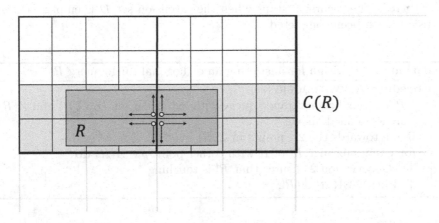

**Fig. 6.** The figure illustrates a range $R$ together with its canonical box $C(R)$. The four points in the center of $C(R)$ are $Z(R)$. The midpoint between these points is $z$, the center of the canonical box $C(R)$.

Every $R_i$ contains exactly one point of $Z(R)$. The points in $Z(R)$ are corner points of the subranges $R_i$. See Fig. 6 for a visualization of $R$, $C(R)$, $Z(R)$ and $z$.

To show that we can answer the subranges efficiently, we need another property. Fix a value $q \in [m]$ and a dimension $i \in D$. Let $H(q, i) = \{x \in M(m, d) \mid x_i = q\}$. The following observation is crucial for the core of our algorithm:

**Observation 3.** *Let $R$ be touching. Then $H(q, i) \cap R$ is also touching. Its extension set is $D \setminus \{i\}$.*

## 3.1   The Algorithm

Our algorithm for answering a range works in two main steps. The first step splits the range into the $2^{d'}$ subranges as defined above. Then each subrange is answered separately by a recursive algorithm. The base case of this recursion answers queries for touching lines $L$ using $O(d \cdot |L|)$ hops. In our description and the analysis, we only consider the processing until all points of $R$ have been visited by our algorithm. For answering the range, these points must be sent back to the querying peer. However, this can easily be achieved by reversing the steps our algorithm does implying only a constant factor of two on our bounds.

To answer a range query, Algorithm 1 is called. Initially, $D$, $C(R)$, $Z(R)$ and the $2^{d'}$ subranges of $R$ are determined. Then, the algorithm routes towards the points of $Z(R)$ and answers the touching subranges independently of each other. Consider such a subrange. Algorithm 2 requires a touching range $R$ with extension set $D$ and a corner point $s = (s_1, \ldots, s_d)$ as input. Given this input, an arbitrary dimension $i \in D$ is selected. The algorithm queries a touching line $L$ with extension $i$ that crosses $R$. The reporting of the points on $L$ is interrupted after the visit of each point $q = (q_1, \ldots, q_{d'})$ of $L$. Now a recursive call for the $H(q_i, i) \cap R$ with extension set $D \setminus \{i\}$ (c.f. Observation 3) and corner point $q$ is triggered. The recursion stops when the extension set $D$ is empty, i.e., all dimensions have been completed.

---

**Algorithm 1.** Algorithm for answering an orthogonal range query $R$

---

1: **procedure** ANSWERQUERY($R$)

   \* $R$ is an orthogonal range query with extension set $D$. $C(R)$ and $Z(R)$ are as defined above. \*

2:    Route towards the $2^{d'}$ points in $Z(R)$

3:    **for each** subrange $R'$ of $R$ with corner point $p \in Z(R)$ **do**

      \* Observation 2 ensures that $R'$ is touching \*

4:       PROCESSRANGE($R'$, $p$)

---

**Algorithm 2.** Algorithm for answering a touching $R$ beginning at a corner $s$

---
1: **procedure** PROCESSRANGE($R$, $s = (s_1, \ldots, s_d)$)
       * Let $D$ be the extension set of $R$ *
2:    **if** $D \neq \emptyset$ **then**
3:       $i \leftarrow$ arbitrary dimension in $D$
4:       $L \leftarrow$ line with extension $i$ and endpoint $s$ that crosses $R$
5:       Visit each point of $L$ consecutively
6:       **for each** visited point $q = (q_1, \ldots, q_{d'})$ on $L$ **do**
         * Due to Observation 3, $H(q_i, i)) \cap R$ is touching *
7:          PROCESSRANGE($H(q_i, i) \cap R$, $q$)

---

## 3.2 Analysis

For the analysis, we are interested in the total number of hops as well as the number of communication rounds. The total number of hops reflects the message complexity of our solution. Our main result is stated in Theorem 1.

**Theorem 1.** *Algorithm 1 answers a range query $R$ in $\mathcal{O}\left(2^{d'} d \log m + d |R|\right)$ hops within $\mathcal{O}\left(d \log m + d \sum_{i=1}^{d}(b_i - a_i + 1)\right)$ communication rounds.*

We already discussed that $|Z(R)| = 2^{d'}$. For the points in $Z(R)$, our algorithm has to do large routing steps that cost $\mathcal{O}(d \log m)$ hops each by Observation 1. Therefore, the first part of our algorithm requires $\mathcal{O}(2^{d'} d \log m)$ hops in total. The subranges can be answered in parallel such that the first part needs $\mathcal{O}(d \log m)$ communication rounds leading to the correctness of Lemma 2.

**Lemma 2.** *Algorithm 1 needs $\mathcal{O}\left(2^{d'} d \log m\right)$ hops and $\mathcal{O}(d \log m)$ communication rounds to route to all points in $Z(R)$.*

It is left to analyze the performance of Algorithm 2 for a touching subrange $R'$ with extension set $D$, and the corner point $c \in Z(R)$ of $R'$. The recursive formulation reduces the problem of answering $R'$ to the problem of answering touching lines crossing $R'$. Due to Observations 2 and 3, we know that all queried lines are touching. Thus, we show Lemma 3. Extending the result of Lemma 3 to the behavior of Algorithm 2 allows to show Lemma 4.

**Lemma 3.** *Let $L$ be a touching line query. $L$ can be answered in $\mathcal{O}(d |L|)$ hops if the routing starts at an endpoint of $L$.*

*Proof.* We observe that due to the touching property of $L$, the smallest surrounding canonical submesh $S$ has a side-length $t$ with $|L| \leq t \leq 2|L|$. The number of hops for visiting $L$ is at most the number of hops for visiting $L'$ which is $L$ extended to cross $S$. $L'$ has length $t$. Observe that $L'$ lies completely in two canonical submeshes $S'$ and $S''$, each of edge-length $t/2$. To analyze the number of hops $F(t)$ for visiting $L'$, we consider the parts of $L'$ in $S'$ and $S''$ separately.

Then, $F(t)$ is composed of $2\,F(t/2)$ for answering the two parts of $L'$, plus the number of hops for jumping from the endpoint of the line in $S'$ to a neighboring endpoint of the line in $S''$. Since $L'$ is completely contained in $S$, Observation 1 implies that the number of hops for the jump is in $\mathcal{O}\,(d \cdot \log t + 1)$. This yields the following recursion:

$$F(t) \le 2 \cdot F\left(\frac{t}{2}\right) + c \cdot d \cdot \log(t) + 1 \qquad \text{for } t > 1$$
$$F(1) = 0 \qquad \text{else}$$

The solution for this recursion is $F(t) = (t-1)\,(2\,cd+1) - cd\,\log(t)$. As $t = |L'| \ge |L|$ this results in a number of hops of $2\,cd\,|L| \in O(d \cdot |L|)$ for answering the line query $L$. □

**Lemma 4.** *Let $R'$ be a subrange of $R$ with extensions $D$. Algorithm 2 answers $R'$ in $\mathcal{O}\,(d\,|R'|)$ hops and $\mathcal{O}\left(d\sum_{i=1}^{d}(b_i - a_i + 1)\right)$ communication rounds when starting at a corner point of $R'$.*

*Proof.* Let $p$ be the corner point of $R'$ at which the routing starts. Due to Observation 2, we know that any line that crosses $R'$ is touching and can be answered efficiently as captured in Lemma 3. Our algorithm consecutively fixes the dimensions in $D$. For each fixed dimension, a touching line $L$ is answered. Each such line $L$ can be answered due to Lemma 3 in $\mathcal{O}\,(d\log|L|)$ hops.

Let $(1,\ldots,d')$ be the sequence of dimensions of $D$ which are fixed one by one by Algorithm 2. Let $r_i = b_i - a_i + 1$ be the side-length of $R'$ along dimension $i$. Then $|R'| = \prod_{i=1}^{d'} r_i$. The number of hops $T(r_1,\ldots,r_{d'})$ needed to report the (touching) range $R'$ is bounded by the following recursion:

$$T(r_1,\ldots,r_{d'}) \le c\,d\,r_1 + r_1\,T(r_2,\ldots,r_{d'}) \qquad \text{for } d' > 1$$
$$T(r_1) \le c\,d\,r_1 \qquad \text{for } d' = 1$$

for a sufficiently large constant $c$. Thus, $T(r_1,\ldots,r_{d'}) \le c\,d\,(r_1 + r_1\,r_2 + \cdots + r_1 \cdot r_2 \cdot \ldots \cdot r_{d'}) \le 2\,c\,d\,r_1\,r_2\,\ldots\,r_{d'} = 2\,cd\,|R'|$. Therefore, all points can be visited in $\mathcal{O}\,(d\,|R'|)$ hops.

Note that the recursive steps for every point of a line can be processed in parallel. Therefore, the algorithm needs $\mathcal{O}\left(d\sum_{i=1}^{d}(b_i - a_i + 1)\right)$ communication rounds. □

Combining Lemmas 2 and 4, we obtain the bounds of Theorem 1.

## 4   Concluding Remarks

It is easy to scale down our construction to smaller P2P networks: for some $m' < m$ let each peer take care of a whole canonical submesh with edge length $m'$. Then only $\left(\frac{m}{m'}\right)^d$ peers are used.

A more interesting question is how to deal with sparse data sets: If only a few of the $m^d$ points of $M(m, d)$ hold data records, we would like to achieve a hop count close to the number of records contained in the queried range. For this, it is interesting to investigate whether our combination of the Hilbert Curve and the Distance Halving Network can be extended to incorporate advantages of, for example, k-d trees.

# References

1. Alber, J., Niedermeier, R.: On multidimensional curves with hilbert property. Theory Comput. Syst. **33**(4), 295–312 (2000). https://doi.org/10.1007/s002240010003
2. Albrecht, J., Oppenheimer, D., Vahdat, A., Patterson, D.A.: Design and implementation trade-offs for wide-area resource discovery. ACM Trans. Internet Technol. **8**(4), 18:1–18:44 (2008). https://doi.org/10.1145/1391949.1391952
3. Andrzejak, A., Xu, Z.: Scalable, efficient range queries for grid information services. In: P2P 2002 Proceedings of the Second International Conference on Peer-to-Peer Computing, pp. 33–40 (2002). https://doi.org/10.1109/PTP.2002.1046310
4. Cai, M., Frank, M., Chen, J., Szekely, P.: MAAN: a multi-attribute addressable network for grid information services. J. Grid Comput. **2**(1), 3–14 (2003). https://doi.org/10.1007/s10723-004-1184-y
5. Carlini, E., Lulli, A., Ricci, L.: DRAGON: multidimensional range queries on distributed aggregation trees. Future Gener. Comput. Syst. **55**, 101–115 (2016). https://doi.org/10.1016/j.future.2015.07.020, http://www.sciencedirect.com/science/article/pii/S0167739X15002526
6. Datta, A., Hauswirth, M., John, R., Schmidt, R., Aberer, K.: Range queries in trie-structured overlays. In: P2P 2005 Proceedings of the Fifth IEEE International Conference on Peer-to-Peer Computing, pp. 57–66. IEEE (2005)
7. Ganesan, P., Yang, B., Garcia-Molina, H.: One torus to rule them all: multidimensional queries in P2P systems. In: WebDB 2004 Proceedings of the 7th International Workshop on the Web and Databases: Colocated with ACM SIGMOD/PODS 2004 WebDB 2004, pp. 19–24. ACM, New York (2004). https://doi.org/10.1145/1017074.1017081
8. Hilbert, D.: Über die stetige Abbildung einer Linie auf ein Flächenstück, pp. 1–2. Springer, Heidelberg (1935). https://doi.org/10.1007/978-3-662-38452-7-1
9. Jagadish, H.V., Ooi, B.C., Vu, Q.H.: Baton: a balanced tree structure for peer-to-peer networks. In: Proceedings of the 31st International Conference on Very Large Data Bases VLDB Endowment, pp. 661–672 (2005)
10. Lee, K., Choi, T., Boykin, P.O., Figueiredo, R.J.: MatchTree: flexible, scalable, and fault-tolerant wide-area resource discovery with distributed matchmaking and aggregation. Future Gener. Comput. Syst. **29**(6), 1596–1610 (2013). https://doi.org/10.1016/j.future.2012.08.009, http://www.sciencedirect.com/science/article/pii/S0167739X12001653. including Special sections: High Performance Computing in the Cloud & Resource Discovery Mechanisms for P2P Systems
11. Li, D., Cao, J., Lu, X., Chen, K.C.C.: Efficient range query processing in peer-to-peer systems. IEEE Trans. Knowl. Data Eng. **21**(1), 78–91 (2009). https://doi.org/10.1109/TKDE.2008.99
12. Kleinberg, J.M.: Navigation in a small world. Nature **406**, 845 (2000)
13. Naor, M., Wieder, U.: Novel architectures for P2P applications: the continuous-discrete approach. ACM Trans. Algorithms **3**(3), 34 (2007). https://doi.org/10.1145/1273340.1273350

14. Pitoura, T., Ntarmos, N., Triantafillou, P.: Saturn: range queries, load balancing and fault tolerance in DHT data systems. IEEE Trans. Knowl. Data Eng. **24**(7), 1313–1327 (2012). https://doi.org/10.1109/TKDE.2010.266
15. Ramabhadran, S., Ratnasamy, S., Hellerstein, J.M., Shenker, S.: Prefix hash tree: an indexing data structure over distributed hash tables. In: Proceedings of the 23rd ACM Symposium on Principles of Distributed Computing, January 2004
16. Schmidt, C., Parashar, M.: Squid: enabling search in DHT-based systems. J. Parallel Distrib. Comput. **68**, 962–975 (2008)
17. Shen, H., Xu, C.Z.: Leveraging a compound graph-based DHT for multi-attribute range queries with performance analysis. IEEE Trans. Comput. **61**(4), 433–447 (2012). https://doi.org/10.1109/TC.2011.30
18. Shu, Y., Ooi, B.C., Tan, K.L., Zhou, A.: Supporting multi-dimensional range queries in peer-to-peer systems. In: P2P 2005 Proceedings of the Fifth IEEE International Conference on Peer-to-Peer Computing, pp. 173–180, August 2005. https://doi.org/10.1109/P2P.2005.35
19. Stoica, I., et al.: Chord: a scalable peer-to-peer lookup protocol for internet applications. IEEE/ACM Trans. Network. **11**(1), 17–32 (2003)
20. Zhang, C., Krishnamurthy, A., Wang, R.Y.: Skipindex: Towards a scalable peer-to-peer index service for high dimensional data. Technical report, Princeton University, May 2004

# A Fully Polynomial Time Approximation Scheme for Packing While Traveling

Frank Neumann[1]($\boxtimes$), Sergey Polyakovskiy[2], Martin Skutella[3], Leen Stougie[4], and Junhua Wu[1]

[1] Optimization and Logistics, School of Computer Science,
The University of Adelaide, Adelaide, Australia
Frank.neumann@adelaide.edu.au
[2] School of Information Technology, Deakin University, Geelong, Australia
[3] Institut für Mathematik, Technische Universität Berlin, Berlin, Germany
[4] CWI, INRIA-Erable and Department of Econometrics and Operations Research,
Vrije Universiteit, Amsterdam, The Netherlands

**Abstract.** Understanding the interaction between different combinatorial optimization problems is a challenging task of high relevance for numerous real-world applications including modern computer and memory architectures as well as high performance computing. Recently, the Traveling Thief Problem (TTP), as a combination of the classical traveling salesperson problem and the knapsack problem, has been introduced to study these interactions in a systematic way. We investigate the underlying non-linear Packing While Traveling Problem (PWTP) of the TTP where items have to be selected along a fixed route. We give an exact dynamic programming approach for this problem and a fully polynomial time approximation scheme (FPTAS) when maximizing the benefit that can be gained over the baseline travel cost. Our experimental investigations show that our new approaches outperform current state-of-the-art approaches on a wide range of benchmark instances.

## 1 Introduction

Combinatorial optimization problems play a crucial role in diverse application areas such as planning, scheduling, and routing, as well as for the efficient use of modern cloud-based computer architectures as well as high performance computing. Many combinatorial optimization problems have been studied extensively in the literature. Two of the most prominent ones are the traveling salesperson problem (TSP) and the knapsack problem (KP). Numerous high performing algorithms have been designed for these two problems.

Looking at combinatorial optimization problems arising in real-world applications, one can observe that real-world problems often are composed of different types of combinatorial problems. For example, delivery problems usually consists of a routing part for the vehicle(s) and a packing part of the goods onto the vehicle(s). Recently, the Traveling Thief Problem (TTP) [1] has been introduced

© Springer Nature Switzerland AG 2019
Y. Disser and V. S. Verykios (Eds.): ALGOCLOUD 2018, LNCS 11409, pp. 59–72, 2019.
https://doi.org/10.1007/978-3-030-19759-9_5

to study the interactions of different combinatorial optimization problems in a systematic way and to gain better insights into the design of multi-component problems. The TTP combines the TSP and KP by making the speed that a vehicle travels along a TSP tour dependent on the weight of the already selected items. Furthermore, the overall objective is given by the sum of the profits of the collected items minus the weight dependent travel cost along the chosen route. A wide range of heuristic search algorithms [2,3,8] and a large benchmark set [12] have been introduced for the TTP in recent years. However, up to now there are no high performing exact approaches to deal with the TTP. On the other hand, the study of non-linear planning problems is an important topic and the design of efficient approximation algorithms has gained increasing interest in recent years [6,15].

The non-linear Packing While Traveling Problem (PWTP) has been introduced in [13] to push forward systematic studies on multi-component problems and deals with the packing part combined with the non-linear travel cost function of the TTP. The PWTP can be seen as the TTP when the route is fixed but the cost still depends on the weight of the items on the vehicle.

**Problem Definition.** The PWTP is formally defined as follows. Given are $n$ cities $1, \ldots, n$, distances $d_i \geq 0$, $1 \leq i \leq n-1$, from city $i$ to city $i+1$, together with $n$ items, one at each city. The item at city $i$ has a non-negative integer profit $p_i$ and weight $w_i$. A vehicle of capacity $W$ travels through the cities in the given order $1, \ldots, n$, and can collect any subset of items $S \subseteq \{1, \ldots, n\}$ of total weight $w(S) := \sum_{i \in S} w_i \leq W$. When traveling from city $i$ to city $i+1$, the speed $v$ of the vehicle depends on the total weight of so far collected items $S_i := S \cap \{1, \ldots, i\}$. More precisely, its speed is an affine linear function of the weight $k = w(S_i)$ given by

$$v(k) := v_{\max} + \frac{k}{W}(v_{\min} - v_{\max}), \tag{1}$$

where $v_{\max}$ is the given maximum possible speed (for the unloaded vehicle) and $v_{\min}$ the given minimum speed (for the fully loaded vehicle). The time $t_i(S_i)$ to travel from city $i$ to city $i+1$ is thus equal to the distance $d_i$ divided by the speed $v_i(w(S_i))$. The objective is to choose a subset of items $S \subseteq \{1, \ldots, n\}$ that maximizes the total benefit $b(S) := p(S) - t(S)$, where $p(S) = \sum_{i \in S} p_i$ is the total profit of selected items and $t(S) := \sum_{i=1}^{n-1} t_i(S_i)$ is the total travel time.

In a slightly more general version of the PWTP, there may be several items or no item at any city $i$. Notice, however, that this can be easily reduced to the special case introduced above. A city with $k > 1$ items can be split into a subsequence of $k$ cities with distances 0 between them. Moreover, at a city with no item we may place a dummy item of profit and weight zero.[1] Further generalizations and interesting variants of the PWTP include other models of weight-dependent travel times occurring in a variety of different application contexts discussed below that can also be handled by the algorithmic techniques introduced in this paper.

---

[1] Alternatively, an intermediate city with no item might be deleted from the sequence.

The PWTP is $NP$-hard even without the capacity constraint usually imposed on the knapsack. Furthermore, exact and approximate mixed integer programming approaches as well as a branch-infer-and-bound approach [11] have been developed for this problem.

**Applications.** The Packing While Traveling Problem is originally motivated by gaining advanced precision when minimizing transportation costs that may have non-linear nature, for example, in applications where weight impacts the fuel costs [4,7]. From this point of view, the problem is a baseline problem in various vehicle routing problems with non-linear costs. Some specific applications of the PWTP may deal with a single truck collecting goods in large remote areas without alternative routes, that is, there may exist a single main route that a vehicle has to follow while any deviations from it in order to visit particular cities are negligible [11].

Applications in the area of modern computing systems include the collection and processing of data by streaming algorithms [16]. Here the sequence of cities/items $1, \ldots, n$ corresponds to a data stream and the capacity $W$ models a bound on the available memory. For multi-level memory architectures, the PWTP's weight-dependent 'travel times' can be interpreted as data processing and computing times that increase with higher memory load; see, e.g., [9]. Further applications in this context include the efficient processing of large amounts of data in social networks and related contexts.

**Our Contribution.** We introduce a dynamic programming approach for the PWTP. The key idea is to consider the items in the order $1, \ldots, n$ they appear along the route that needs to be traveled and apply dynamic programming similar as for the classical knapsack problem [14]. When considering an item, the decision has to be made on whether or not to pack the item. The dynamic programming approach computes for the first $i$ items, $1 \leq i \leq n$, and possible subsets of weight $\bar{w}$ the maximal objective value that can be obtained. As the programming table that is used depends on the number of different possible weights, the algorithm runs in pseudo-polynomial time.

After having obtained the exact approach based on dynamic programming, we consider the design of a fully polynomial approximation scheme (FPTAS) [5]. First, we show that it is $NP$-hard to decide whether a given instance of the PWTP has a non-negative objective value. This rules out any polynomial time algorithm with finite approximation ratio, unless $P = NP$. Due to this, we design an FPTAS for the amount that can be gained over the travel cost when the vehicle travels empty (which is the minimal possible travel cost). Our FPTAS is based on the observation that the item with the largest benefit leads to an objective value of at least $OPT/n$ and uses appropriate rounding in the previously designed dynamic programming approach. An interesting and distinguishing feature of our FPTAS is the fact that, in contrast to the standard approach in the area of approximation schemes, we do not explicitly round values to arrive at a polynomial-size state space of the dynamic program. Instead, an approximate domination criterion is used to restrict to a polynomial number of intermediate states.

We evaluate our two approaches on a wide range of instances from the TTP benchmark set [12], and compare them to the exact and approximative approaches given in [11]. Our results show that the large majority of the instances that can be handled by exact methods, are solved much faster by dynamic programming than the previously developed mixed integer programming and branch-infer-and-bound approaches. Considering instances with a larger profit and weight range, we show that the choice of the approximation guarantee significantly impacts the runtime behavior.

**Outline.** The paper is structured as follows. In Sect. 2 we present the exact dynamic programming approach, and design an FPTAS in Sect. 3. Our experimental results are discussed in Sect. 4. Finally, we finish with some conclusions.

## 2   Dynamic Programming

We introduce a dynamic programming approach for solving the PWTP. Dynamic programming is one of the traditional approaches for the classical knapsack problem [14]. The dynamic programming table $\beta$ consists of $n$ rows, indexed by $i = 1, \ldots, n$, and $W + 1$ columns, indexed by $k = 0, \ldots, W$. Items are processed in the order $i = 1, \ldots, n$ they appear along the tour. The entry $\beta(i, k)$ shall denote the maximal benefit that can be obtained by considering all subsets of the first $i$ items $\{1, \ldots, i\}$ of total weight exactly $k$, for $k = 0, \ldots, W$. We denote by $\beta(i, \cdot)$ the row containing the entries $\beta_{i,k}$. In the case that a subset of total weight $k$ does not exist, we set $\beta(i, k) := -\infty$.

Let $d_{i,n} := \sum_{j=i}^{n-1} d_j$ be the distance from city $i$ to the last city $n$. We denote by $b(\emptyset) := -d_{1,n}/v_{\max}$ the benefit of the empty set, that is, the travel cost when the vehicle travels empty. Furthermore, the benefit when only item $i$ is chosen is

$$b(\{i\}) := b(\emptyset) + p_i - \frac{d_{i,n}}{v(w_i)} + \frac{d_{i,n}}{v_{\max}},$$

as the vehicle will now only travel at speed $v(w_i)$ from city $i$ on. The entries in the first row can be easily computed as

$$\beta(1, k) := \begin{cases} b(\emptyset) & \text{if } k = 0 \neq w_1, \\ b(\{1\}) & \text{if } k = w_1, \\ -\infty & \text{otherwise.} \end{cases} \tag{2}$$

For $i = 2, \ldots, n$, based on the row $\beta(i-1, \cdot)$ we can compute the next row $\beta(i, \cdot)$. To keep notation simple, we let $\beta(i-1, q) := -\infty$ for $q < 0$. Then,

$$\beta(i, k) := \max \left\{ \beta(i-1, k), \beta(i-1, k-w_i) + p_i - \frac{d_{i,n}}{v(k)} + \frac{d_{i,n}}{v(k-w_i)} \right\}. \tag{3}$$

The correctness of this recursive formula is discussed in the proof of the next theorem.

**Theorem 1.** *For each $i$ and $k$, the entry $\beta(i, k)$ stores the maximal possible benefit $b(S)$ over all subsets $S$ of $\{1, \ldots, i\}$ having weight exactly $k$. In particular, $\max_k \beta(n, k)$ is the value of an optimal solution, which can be obtained via backtracking.*

*Proof.* We use induction on $i$. The statement is true for $i = 1$ as there are only the two options of choosing or not choosing the first item, which are both considered in (2). Now assume that $\beta(i - 1, k)$ stores the maximal benefit for each weight $k$ when considering all subsets of $\{1, \ldots, i - 1\}$. Notice that for a subset $S' \subseteq \{1, \ldots, i-1\}$ of weight at most $W - w_i$, the benefit of $S' \cup \{i\}$ equals

$$ b(S' \cup \{i\}) = b(S') + p_i - \left( \frac{d_{i,n}}{v\big(w(S') + w_i\big)} - \frac{d_{i,n}}{v\big(w(S')\big)} \right), \qquad (4) $$

since adding item $i$ to subset $S'$ leads to the reduced speed $v\big(w(S') + w_i\big)$ of the vehicle, instead of $v\big(w(S')\big)$, from city $i$ on. Consider now a subset $S \subseteq \{1, \ldots, i\}$ with $w(S) = k$ of maximum benefit $b(S)$. If $i \notin S$, then $S$ must obviously be a maximum benefit subset of $\{1, \ldots, i - 1\}$ of weight $k$ as well. In particular, $b(S) = \beta(i - 1, k)$; see the first term on the right-hand side of (3). Otherwise, if $i \in S$, then $S = S' \cup \{i\}$ for a maximum benefit subset $S' \subseteq \{1, \ldots, i - 1\}$ of weight $k - w_i$, that is, $b(S') = \beta(i - 1, k - w_i)$. Notice that the second term on the right-hand side of (3) thus coincides with (4). This concludes the proof.

Finally, we investigate the runtime for this dynamic program. If $d_{i,n}$ has been computed for each $i$, which takes $O(n)$ time in total, then each entry of the dynamic programming table $\beta$ can be computed in constant time. Thus, the running time of the dynamic program is in $O(nW)$. To empirically speed up the computation of the dynamic program, it is sufficient to only store an entry for $\beta(i, k)$ if it is not dominated by any other entry in $\beta(i, \cdot)$, that is, if there is no $k' < k$ with $\beta(i, k') \geq \beta(i, k)$. This is justified by the following lemma.

**Lemma 1.** *The increase in travel cost due to a new item $i$ given by the term in brackets on the right-hand side of (4) is an increasing function of the weight $w(S')$ of so far collected items.*

*Proof.* For $v(k)$ as defined in (1), let $t(k) := 1/v(k)$ denote the travel time per unit distance when the vehicle has collected items of total weight $k$. Notice that the thereby defined function $t : [0, W] \to \mathbb{R}_{\geq 0}$ is convex and increasing.

## 3   Approximation Algorithms

We now turn our attention to approximation algorithms. The NP-hardness proof for the PWTP given in [11] does not rule out polynomial time approximation algorithms. In this section, we first show that polynomial time approximation algorithms with a finite approximation ratio do not exist, unless $P = NP$. This results motivates the design of an FPTAS for the shifted objective function given by the amount that can be gained over the baseline cost when the vehicle is traveling empty.

## 3.1   Inapproximability of the Packing While Traveling Problem

The objective function for PWTP can take on positive and negative values. We show that deciding whether a given PWTP instance has a solution that is non-negative is already NP-complete.

**Theorem 2.** *Given a PWTP instance, the problem to decide whether there is a solution $S \subseteq \{1, \ldots, n\}$ with $b(S) \geq 0$ is NP-complete.*

*Proof.* The problem is obviously in NP as one can verify in polynomial time for a given solution $S$ whether $b(S) \geq 0$ holds by evaluating the objective function. It remains to show that the problem is NP-hard.

We reduce the $NP$-complete *Subset Sum Problem* (SSP) to our problem. An instance of SSP is given by $n$ positive integers $\{s_1, \ldots, s_n\}$ and a positive integer $Q$. The question is whether there exists a subset $S \subseteq \{1, \ldots, n\}$ such that $\sum_{i \in S} s_i = Q$. Given an instance of SSP, we construct an instance of PWTP consisting of $n$ cities and items of profit and weight $p_i = w_i = s_i$, for $i = 1, \ldots, n$. The distances $d_i$ between cities are all equal to zero except for the last distance $d_{n-1} := Q^2$. Finally, the vehicle has capacity $W := Q$ and its minimum and maximum speed are $v_{\min} := v_{\max} := Q$, that is, the speed does not depend on the weight of collected items. It is easy to see that the benefit of any solution $S \subseteq \{1, \ldots, n\}$ is equal to $b(S) = p(S) - Q = \sum_{i \in S} s_i - Q$. In particular, as $p(S) = w(S) \leq W = Q$, it holds that $b(S) \geq 0$ if and only if $S$ is a solution to the underlying instance of the SSP.

We can even prove the following slightly stronger complexity result.

**Proposition 1.** *The decision version of the PWTP stated in Theorem 2 is even NP-hard if the vehicle capacity is large enough to fit all items, that is, if $W \geq w(\{1, \ldots, n\})$.*

*Proof.* We modify the reduction given in the proof of Theorem 2 as follows. First of all we restrict to instances of the SSP with $\sum_{i=1}^{n} s_i = 2Q$ (in other words, we give a reduction from the NP-complete Partition Problem). The vehicle capacity is then set to $W := 2Q$, the maximum speed to $v_{\max} := 2Q$, and the minimum speed to $v_{\min} := 0$. Then, the benefit of a subset of items $S \subseteq \{1, \ldots, n\}$ is

$$b(S) = p(S) - \frac{Q^2}{2Q - w(S)} = w(S) - \frac{Q^2}{2Q - w(S)}.$$

We consider the right-hand side term as a function of $w(S)$. It is easy to check that this function attains its unique maximum of value 0 for $w(S) = Q$.

As a corollary of Theorem 2, we obtain the following non-approximability result.

**Corollary 1.** *There is no polynomial time approximation algorithm for PWTP with a finite approximation ratio, unless $P = NP$.*

## 3.2   An FPTAS for Amount over Baseline Travel Cost

In view of Corollary 1, we shift the objective function value and consider the amount that can be gained over the cost when the vehicle travels empty as the new objective. More precisely, for a subset of items $S \subseteq \{1, \ldots, n\}$ the new objective is

$$b'(S) := b(S) - b(\emptyset).$$

This is motivated by the scenario where the vehicle has to travel along the given route anyway, and the goal is to maximize the gain over this (negative) baseline cost $b(\emptyset)$. Notice that an optimal solution for this objective is also an optimal solution for the original PWTP objective. Approximation results, however, do not carry over as the objective value is shifted by $b(\emptyset)$.

As in the proof of Lemma 1, let $t(k)$ be the travel time per unit distance when the vehicle has collected items of total weight $k$. It follows from the proof of Lemma 1 that, for each item $i$ and $0 \leq k \leq W - w_i$, we get

$$t(k + w_i) - t(k) \geq t(w_i) - t(0).$$

This means that the marginal cost (with respect to the travel time) of adding an item is lowest if there are no other items chosen. As a consequence, we get for each subset $S \subseteq \{1, \ldots, n\}$ with $w(S) \leq W$ that

$$b'(S) \leq \sum_{i \in S} b'(\{i\}).$$

In particular, when choosing an optimal subset $S$ maximizing $b'(S) =: \mathrm{OPT}$, there is an $i \in S$ with $b'(i) \geq \mathrm{OPT}/|S| \geq \mathrm{OPT}/n$. Thus, $L := \max_{1 \leq i \leq n} b'(\{i\})$ provides an efficiently computable lower bound on the value of an optimal solution satisfying $\mathrm{OPT}/n \leq L \leq \mathrm{OPT}$.

In order to obtain a fully polynomial time approximation scheme (FPTAS) for the problem of maximizing $b'(S)$ over all feasible subsets $S \subseteq \{1, \ldots, n\}$, we start by carefully modifying the dynamic programming scheme from Sect. 2 given by Eqs. (2) and (3) as follows. Let

$$\beta'(1, k) := \begin{cases} b'(\emptyset) & \text{if } k = 0 \neq w_1, \\ b'(\{1\}) & \text{if } k = w_1, \\ -\infty & \text{otherwise.} \end{cases}$$

Then, for $i = 2, \ldots, n$, let

$$\beta'(i, k) := \max\left\{\beta'(i - 1, k), \beta'(i - 1, k - w_i) + p_i - \frac{d_{i,n}}{v(k)} + \frac{d_{i,n}}{v(k - w_i)}\right\}.$$

As discussed at the end of Sect. 2, we can speed up the dynamic program by setting $\beta'(i, k) := -\infty$ in case there is a $k' < k$ with $\beta'(i, k') \geq \beta'(i, k)$.

The idea of the FPTAS described in Algorithm 1 is to further speed up the dynamic program by ignoring entries $\beta'(i, k)$ such that there is a $k' < k$ with

---

**Algorithm 1.** FPTAS for maximizing $b'(S)$

1. set $L := \max_{1 \leq i \leq n} b'(\{i\})$, $r := \epsilon L/n$, and $d_{i,n} := \sum_{j=i}^{n-1} d_j$ for $1 \leq i \leq n$;
2. initially, all values $\beta(i,k)$ are assumed to be $-\infty$;
3. set $\beta'(1,0) := b'(\emptyset)$ and $\beta'(1,w_1) := b'(\{1\})$;
4. for $i = 1, \ldots, n - 1$ do:
5.    for each $k$ with $\lfloor \beta'(i,k)/r \rfloor > \max\{\lfloor \beta'(i,k')/r \rfloor, -\infty\}$ for all $k' < k$ do:
6.       set $\beta'(i+1,k) := \max\{\beta'(i,k), \beta'(i+1,k)\}$;
7.       if $k^+ := k + w_{i+1} \leq W$, set

$$\beta'(i+1,k^+) := \max\{\beta'(i,k) + p_{i+1} - \frac{d_{i+1,n}}{v(k^+)} + \frac{d_{i+1,n}}{v(k)}, \beta'(i+1,k^+)\}$$

8. determine $\max_k \beta'(n,k)$ and corresponding solution $S$ by backtracking;

---

$\lfloor \beta'(i,k)/r \rfloor > \lfloor \beta'(i,k')/r \rfloor$ for $r := \epsilon L/n$. Due to this, in terms of the objective function we lose at most $r$ in every row of the dynamic programming table. The overall loss is thus bounded by $nr = \epsilon L \leq \epsilon \mathrm{OPT}$.

**Theorem 3.** *Algorithm 1 is an FPTAS for the problem to maximize $b'(S)$ over all subsets of items $S \subseteq \{1, \ldots, n\}$ with $w(S) \leq W$.*

*Proof.* As argued above, the value of the computed solution is at least $(1 - \epsilon)\mathrm{OPT}$. It remains to argue that the running time of Algorithm 1 is bounded by a polynomial in the input size and $1/\epsilon$. This can be seen as follows:

**Claim.** For the dynamic programming table $\beta'$ computed by Algorithm 1, there are at most $O(n^2/\epsilon)$ entries of finite value in row $\beta'(i, \cdot)$, for $i = 1, \ldots, n$.

*Proof of the Claim:* We use induction on $i$. The case $i = 1$ is clear by Step 3 of Algorithm 1. Moreover, the for-loop in Step 5 considers at most $1 + \mathrm{OPT}/r = 1 + n\mathrm{OPT}/(\epsilon L) \leq 1 + n^2/\epsilon$ different values of $k$. For each such $k$, at most two entries in the next row $i + 1$ are modified. This concludes the proof of the claim. The overall running time is thus polynomial in the input size and $1/\epsilon$.

We conclude this section with the following generalizing remark.

*Remark 1.* The construction of the FPTAS only used the fact that the travel time per unit distance is monotonically increasing and convex. Hence, the FPTAS holds for any PWTP problem where the travel time per unit distance has this property.

## 4    Experiments and Results

In this section, we investigate the effectiveness of the proposed DP and FPTAS approaches based on our implementations in Java. We mainly focus on two issues: (1) studying how the DP and FPTAS perform compared to the state-of-the-art approaches; (2) investigating how the performance and accuracy of the FPTAS change when the parameter $\epsilon$ is altered.

In order to be comparable to the mixed integer programming (MIP) and the branch-infer-and-bound (BIB) approaches presented in [11], we conduct our experiments on the same families of test instances. Our experiments are carried out on a computer with 4 GB RAM and a 3.06 GHz Intel Dual Core processor, which is also the same as the machine used in the paper mentioned above.

We compare the DP to the exact MIP (*eMIP*) and the branch-infer-and-bound approaches as well as the FPTAS to the approximate MIP (*aMIP*), as the former three are all exact approaches and the latter two are all approximations. Table 1 demonstrates the results for a route of 101 cities and various types of packing instances. For this particular family, we consider three types of instances: *uncorrelated* (uncorr), *uncorrelated with similar weights* (uncorr-s-w) and *bounded strongly correlated* (b-s-corr), which are further distinguished by the different correlations between profits and weights. In combination with three different numbers of items and three settings of the capacity, we have 27 instances in total, as shown in the column called "*Instance*". Similarly to the settings in [11], every instance with "_01" postfix has a relatively small capacity. We expect such instances to be potentially easy to solve by DP and FPTAS due to the nature of the algorithms. The *OPT* column shows the optimum of each instance and the *RT(s)* columns illustrate the running time for each of the approaches in the time unit of a second. To demonstrate the quality of an approximate approach applied to the instances, we use the ratio between the objective value obtained by the algorithm and the optimum obtained for an instance as the approximation rate $AR(\%) = 100 \times \frac{OBJ}{OPT}$.

In the comparison of exact approaches, our results show that the DP is much quicker than the exact MIP and BIB in solving the majority of the instances. The exact MIP is slower than the DP in every case and this dominance is mostly significant. For example, it spends around 35 min to solve the instance *uncorr-s-w_10* with 1,000 items, where the DP needs around 15 s only. On the other hand, the BIB slightly beats the DP on three instances, but the DP is superior for the rest 24 instances. An extreme case is *b-s-corr_01* with 1,000 items where the BIB spends above 1.5 h while the DP solves it in 11 s only. Concerning the running time of the DP, it significantly increases only for the instances having large amount of items with strongly correlated weights and profits, such as *b-s-corr_06* and *b-s-corr_10* with 1,000 items. However, *b-s-corr_01* seems exceptional due to the limited capacity assigned to the instance.

Our comparison between the approximation approaches shows that the FPTAS has significant advantages as well. The approximation ratios remain 100% when $\epsilon$ equals 0.0001 and 0.01. Only when $\epsilon$ is set to 0.25, the FPTAS starts to output the results having similar accuracies as the ones of *aMIP*. With regard to the performance, the FPTAS takes less running time than *aMIP* on the majority of the instances despite the setting of $\epsilon$. As an extreme case, *aMIP* requires hours to solve the *uncorr-s-w_01* instance with 1,000 items, but the FPTAS takes less than a second. However, the *aMIP* performs much better on *b-s-corr_06* and *b-s-corr_10* with 1,000 items. This somehow indicates that the underlying factors that make instances hard to solve by approximate MIP and

**Table 1.** Results on small range instances

| Instance | m | OPT | Exact approaches | | | Approximation approaches | | | | | | | | | | | |
|---|---|---|---|---|---|---|---|---|---|---|---|---|---|---|---|---|---|
| | | | eMIP | BIB | DP | aMIP | | FPTAS | | | | | | | | | |
| | | | | | | | | ε = 0.0001 | | ε = 0.01 | | ε = 0.1 | | ε = 0.25 | | ε = 0.75 | |
| | | | RT(s) | RT(s) | RT(s) | AR(%) | RT(s) | AR(%) | RT(s) | AR(%) | RT(s) | AR(%) | RT(s) | AR(%) | RT(s) | AR(%) | RT(s) |
| Instance family eil101 | | | | | | | | | | | | | | | | | |
| uncorr_01 | 100 | 1651.697 | 1.217 | 5.694 | 0.027 | 100 | 3.838 | 100 | 0.001 | 100 | 0.001 | 100 | 0.001 | 100 | 0.001 | 100 | 0.025 |
| uncorr_06 | 100 | 10155.4942 | 12.605 | 3.698 | 0.065 | 100 | 4.961 | 100 | 0.012 | 100 | 0.012 | 100 | 0.011 | 100 | 0.011 | 99.9928 | 0.063 |
| uncorr_10 | 100 | 10297.7134 | 3.525 | 0.795 | 0.036 | 100 | 0.624 | 100 | 0.017 | 100 | 0.017 | 99.9939 | 0.016 | 99.9939 | 0.016 | 99.9653 | 0.037 |
| uncorr-s-w_01 | 100 | 2152.6188 | 0.328 | 7.566 | 0.001 | 100 | 3.978 | 100 | 0 | 100 | 0 | 100 | 0 | 100 | 0 | 100 | 0.003 |
| uncorr-s-w_06 | 100 | 4333.8512 | 12.59 | 2.215 | 0.012 | 100 | 2.699 | 100 | 0.008 | 100 | 0.007 | 100 | 0.007 | 99.9569 | 0.008 | 99.9569 | 0.017 |
| uncorr-s-w_10 | 100 | 9048.4908 | 37.144 | 1.107 | 0.022 | 100 | 1.763 | 100 | 0.012 | 100 | 0.012 | 100 | 0.012 | 100 | 0.012 | 99.9355 | 0.02 |
| b-s-corr_01 | 100 | 4441.9852 | 1.42 | 125.954 | 0.014 | 100 | 5.366 | 100 | 0.01 | 100 | 0.009 | 100 | 0.009 | 100 | 0.009 | 100 | 0.013 |
| b-s-corr_06 | 100 | 10260.9767 | 4.509 | 22.541 | 0.101 | 100 | 2.761 | 100 | 0.058 | 100 | 0.058 | 100 | 0.057 | 100 | 0.048 | 100 | 0.087 |
| b-s-corr_10 | 100 | 13630.6153 | 11.013 | 27.081 | 0.187 | 99.9971 | 3.713 | 100 | 0.103 | 100 | 0.103 | 99.9971 | 0.101 | 99.9606 | 0.081 | 99.8143 | 0.113 |
| uncorr_01 | 500 | 17608.5781 | 19.594 | 27.581 | 0.247 | 100 | 5.757 | 100 | 0.171 | 100 | 0.171 | 100 | 0.161 | 100 | 0.153 | 100 | 0.377 |
| uncorr_06 | 500 | 56294.5239 | 384.213 | 13.354 | 2.829 | 100 | 7.8 | 100 | 2.37 | 100 | 2.344 | 100 | 2.3 | 100 | 2.212 | 100 | 2.34 |
| uncorr_10 | 500 | 66141.484 | 211.302 | 2.325 | 4.01 | 100 | 0.718 | 100 | 3.72 | 100 | 3.645 | 100 | 3.446 | 100 | 3.531 | 100 | 3.632 |
| uncorr-s-w_01 | 500 | 13418.484 | 4.337 | 34.866 | 0.09 | 100 | 50.31 | 100 | 0.085 | 100 | 0.085 | 100 | 0.09 | 100 | 0.084 | 99.991 | 0.085 |
| uncorr-s-w_06 | 500 | 34280.473 | 346.43 | 7.285 | 1.04 | 100 | 9.609 | 100 | 0.964 | 100 | 0.964 | 100 | 0.933 | 100 | 0.905 | 100 | 0.92 |
| uncorr-s-w_10 | 500 | 50836.6558 | 519.902 | 3.338 | 2.022 | 100 | 3.354 | 100 | 2.005 | 100 | 1.783 | 100 | 1.753 | 100 | 1.784 | 100 | 2.147 |
| b-s-corr_01 | 500 | 21306.9158 | 40.482 | 624.204 | 1.534 | 100 | 13.338 | 100 | 1.373 | 100 | 1.279 | 100 | 1.116 | 100 | 0.949 | 100 | 0.716 |
| b-s-corr_06 | 500 | 69370.2367 | 236.387 | 97.313 | 14.616 | 99.9996 | 7.847 | 100 | 13.393 | 100 | 12.975 | 99.9996 | 11.642 | 99.9996 | 9.741 | 99.9996 | 6.018 |
| b-s-corr_10 | 500 | 82033.9452 | 376.569 | 218.728 | 22.011 | 100 | 2.309 | 100 | 21.372 | 100 | 20.829 | 100 | 18.573 | 100 | 15.313 | 99.9943 | 8.84 |
| uncorr_01 | 1000 | 36170.9109 | 218.306 | 114.567 | 1.872 | 99.9993 | 11.918 | 100 | 1.891 | 100 | 1.875 | 99.9993 | 1.832 | 99.9993 | 1.845 | 100 | 1.764 |
| uncorr_06 | 1000 | 93949.1981 | 1261.949 | 36.847 | 20.944 | 100 | 17.971 | 100 | 17.024 | 100 | 16.615 | 100 | 16.545 | 100 | 16.378 | 100 | 15.713 |
| uncorr_10 | 1000 | 122963.6617 | 620.896 | 4.821 | 30.116 | 100 | 2.184 | 100 | 27.305 | 100 | 26.783 | 100 | 26.541 | 100 | 26.051 | 100 | 23.905 |
| uncorr-s-w_01 | 1000 | 27800.9614 | 241.957 | 399.158 | 0.802 | 100 | 4985.566 | 100 | 0.73 | 100 | 0.69 | 100 | 0.688 | 100 | 0.724 | 100 | 0.687 |
| uncorr-s-w_06 | 1000 | 61764.4599 | 1152.624 | 12.792 | 9.872 | 100 | 19.063 | 100 | 8.686 | 100 | 8.812 | 100 | 8.56 | 100 | 8.74 | 100 | 8.396 |
| uncorr-s-w_10 | 1000 | 103572.4074 | 2146.408 | 7.644 | 15.047 | 100 | 9.688 | 100 | 14.03 | 100 | 13.912 | 100 | 13.797 | 100 | 13.982 | 100 | 13.492 |
| b-s-corr_01 | 1000 | 46886.1094 | 378.551 | 6129.531 | 11.783 | 99.9988 | 46.394 | 100 | 11.714 | 100 | 11.358 | 99.9988 | 10.793 | 99.9988 | 9.592 | 99.9988 | 6.536 |
| b-s-corr_06 | 1000 | 125830.6887 | 643.533 | 919.201 | 94.523 | 99.9999 | 10.311 | 100 | 92.411 | 100 | 91.039 | 99.9999 | 83.002 | 99.9999 | 71.078 | 99.9999 | 45.433 |
| b-s-corr_10 | 1000 | 161990.5015 | 862.572 | 1646.52 | 151.601 | 100 | 7.16 | 100 | 150.279 | 100 | 149.722 | 100 | 134.764 | 100 | 113.049 | 99.9981 | 70.135 |

## Table 2. Results of DP and FPTAS on large range instances

| Instance | m | DP OPT | DP RT(s) | FPTAS ε = 0.0001 AR(%) | RT(s) | ε = 0.001 AR(%) | RT(s) | ε = 0.01 AR(%) | RT(s) | ε = 0.1 AR(%) | RT(s) | ε = 0.25 AR(%) | RT(s) | ε = 0.5 AR(%) | RT(s) | ε = 0.75 AR(%) | RT(s) |
|---|---|---|---|---|---|---|---|---|---|---|---|---|---|---|---|---|---|
| **Instance family eil101_large-range** | | | | | | | | | | | | | | | | | |
| uncorr_01 | 100 | 69802802.2801 | 0.03 | 100 | 0.002 | 100 | 0.002 | 100 | 0.002 | 100 | 0.002 | 100 | 0.002 | 100 | 0.002 | 100 | 0.029 |
| uncorr_06 | 100 | 204813765.6933 | 0.053 | 100 | 0.019 | 100 | 0.02 | 100 | 0.019 | 100 | 0.019 | 100 | 0.019 | 100 | 0.019 | 100 | 0.049 |
| uncorr_10 | 100 | 172176182.1249 | 0.041 | 100 | 0.028 | 100 | 0.028 | 100 | 0.028 | 100 | 0.028 | 100 | 0.027 | 100 | 0.026 | 99.9628 | 0.037 |
| uncorr-s-w_01 | 100 | 36420530.5753 | 0.006 | 100 | 0.003 | 100 | 0.003 | 100 | 0.003 | 100 | 0.003 | 100 | 0.003 | 100 | 0.003 | 100 | 0.004 |
| uncorr-s-w_06 | 100 | 148058928.2952 | 0.098 | 100 | 0.072 | 100 | 0.502 | 100 | 0.072 | 100 | 0.072 | 100 | 0.069 | 99.9978 | 0.065 | 99.9978 | 0.07 |
| uncorr-s-w_10 | 100 | 142538516.4602 | 0.136 | 100 | 0.101 | 100 | 0.104 | 100 | 0.104 | 99.9978 | 0.103 | 99.9978 | 0.096 | 99.9978 | 0.086 | 99.9978 | 0.089 |
| m-s-corr_01 | 100 | 19549602.2671 | 0.003 | 100 | 0.002 | 100 | 0.002 | 100 | 0.002 | 100 | 0.002 | 100 | 0.002 | 100 | 0.002 | 100 | 0.002 |
| m-s-corr_06 | 100 | 137203175.1921 | 0.147 | 100 | 0.115 | 100 | 0.118 | 100 | 0.113 | 100 | 0.113 | 100 | 0.089 | 100 | 0.063 | 100 | 0.043 |
| m-s-corr_10 | 100 | 225584278.6004 | 0.424 | 100 | 0.326 | 100 | 0.329 | 100 | 0.312 | 100 | 0.312 | 100 | 0.2 | 100 | 0.179 | 100 | 0.073 |
| uncorr_01 | 500 | 385692662.0930 | 0.47 | 100 | 0.451 | 100 | 0.454 | 100 | 0.619 | 100 | 0.508 | 100 | 0.445 | 100 | 0.43 | 100 | 0.517 |
| uncorr_06 | 500 | 958013934.6172 | 3.539 | 100 | 3.749 | 100 | 7.431 | 100 | 3.947 | 100 | 3.69 | 99.9996 | 3.677 | 99.9996 | 3.486 | 99.9993 | 3.021 |
| uncorr_10 | 500 | 844949838.4389 | 4.87 | 100 | 5.393 | 100 | 5.716 | 100 | 5.483 | 100 | 5.135 | 100 | 4.851 | 99.9992 | 4.609 | 99.9992 | 4.295 |
| uncorr-s-w_01 | 500 | 182418888.9364 | 1.157 | 100 | 1.157 | 100 | 1.199 | 100 | 1.145 | 99.9995 | 1.112 | 99.9995 | 1.063 | 99.9995 | 0.977 | 99.9904 | 0.929 |
| uncorr-s-w_06 | 500 | 780432253.0187 | 22.39 | 100 | 25.04 | 100 | 26.276 | 100 | 24.024 | 100 | 23.282 | 99.9997 | 21.756 | 99.9997 | 18.293 | 99.9997 | 18.411 |
| uncorr-s-w_10 | 500 | 714433353.7957 | 30.959 | 100 | 34.458 | 100 | 39.004 | 100 | 34.308 | 100 | 32.308 | 99.9996 | 28.792 | 99.999 | 26.392 | 99.999 | 25.971 |
| m-s-corr_01 | 500 | 964639441.1275 | 2.335 | 100 | 2.478 | 100 | 2.782 | 100 | 2.695 | 100 | 1.509 | 100 | 0.963 | 100 | 0.546 | 100 | 0.408 |
| m-s-corr_06 | 500 | 666701000.1488 | 108.705 | 100 | 126.833 | 100 | 139.63 | 100 | 122.75 | 100 | 62.479 | 100 | 33.547 | 100 | 17.959 | 100 | 10.642 |
| m-s-corr_10 | 500 | 1082009880.5886 | 262.999 | 100 | 299.862 | 100 | 317.352 | 100 | 274.284 | 100 | 145.087 | 100 | 78.47 | 99.9994 | 41.816 | 99.9994 | 25.924 |
| uncorr_01 | 1000 | 777386336.9660 | 4.222 | 100 | 4.397 | 100 | 4.347 | 100 | 4.309 | 100 | 4.341 | 100 | 4.377 | 100 | 4.28 | 100 | 4.24 |
| uncorr_06 | 1000 | 1933319297.4248 | 46.043 | 100 | 51.383 | 100 | 53.087 | 100 | 48.861 | 100 | 52.957 | 99.9999 | 52.062 | 99.9997 | 50.286 | 99.9996 | 51.488 |
| uncorr_10 | 1000 | 1693797490.1704 | 64.485 | 100 | 76.744 | 100 | 78.847 | 100 | 74.128 | 100 | 82.754 | 100 | 77.057 | 99.999 | 72.283 | 100 | 72.567 |
| uncorr-s-w_01 | 1000 | 361991311.8336 | 14.254 | 100 | 15.072 | 100 | 15.67 | 100 | 14.523 | 100 | 14.11 | 100 | 14.039 | 100 | 12.088 | 100 | 11.129 |
| uncorr-s-w_06 | 1000 | 1574469459.3163 | 286.843 | 100 | 318.096 | 100 | 330.508 | 100 | 337.289 | 100 | 334.318 | 99.9996 | 307.588 | 99.9998 | 270.013 | 99.9996 | 245.927 |
| uncorr-s-w_10 | 1000 | 1439410696.3695 | 393.793 | 100 | 438.775 | 100 | 455.83 | 100 | 464.527 | 100 | 441.955 | 99.9994 | 433.672 | 99.9994 | 378.917 | 99.9994 | 340.813 |
| m-s-corr_01 | 1000 | 191170309.5684 | 46.858 | 100 | 58.031 | 100 | 59.987 | 100 | 58.101 | 100 | 31.703 | 100 | 18.771 | 100 | 10.728 | 100 | 6.831 |
| m-s-corr_06 | 1000 | 1315708161.7720 | 2393.205 | 100 | 2512.281 | 100 | 2606.412 | 100 | 1921.573 | 100 | 666.749 | 100 | 364.452 | 100 | 208.969 | 100 | 150.06 |
| m-s-corr_10 | 1000 | 2163713055.3759 | 6761.49 | 100 | 6668.535 | 100 | 6441.906 | 100 | 4526.653 | 100 | 1334.882 | 100 | 703.258 | 100 | 397.527 | 100 | 282.211 |

FPTAS have different nature. Understanding these factors more and using them wisely should help to build a more powerful algorithm with mixed features of MIP and FPTAS.

In our second experiment, we use test instances which are slightly different to those in the benchmark used in [11]. This is motivated by our findings that relaxing $\epsilon$ from 0.0001 to 0.75 improves the runtime performance of FPTAS by around 50% for the b-s-corr instances, while does not degrade the accuracy noticeably. At the same time, there is no significant improvement for other instances. It's surprising as shows that the performance improvement can be easily achieved on complex instances. Therefore, we study how the FPTAS performs if the instances are more complicated. The idea is to use instances with large weights, which are known to be difficult regarding dynamic programming based approaches for the classical knapsack problem. We follow the same way to create TTP instances as proposed in [12] and generate the knapsack component of the problem as discussed in [10]. Specifically, we extend the range to generate potential profits and weights from $[1, 10^3]$ to $[1, 10^7]$ and focus on *uncorrelated* (uncorr), *uncorrelated with similar weights* (uncorr-s-w), and *multiple strongly correlated* (m-s-corr) types of instances. Additionally, in the stage of assigning the items of a knapsack instance to particular cities of a given TSP tour, we sort the items in descending order of their profits and the second city obtains $k$, $k \in \{1, 5, 10\}$, items of the largest profits, the third city then has the next $k$ items, and so on. We expect that such assignment should force the algorithms to select items in the first cities of a route making the instances more challenging for the DP and FPTAS. In reality, these instances indeed are harder than the ones in the first experiment, which forces us to switch to the 128 GB RAM and $8 \times$ (2.8 GHz AMD 6 core processors) cluster machine to carry out the second experiment.

Table 2 illustrates the results of running the DP and FPTAS on the instances with the large range of profits and weights. Generally speaking, we can observe that the instances are significantly harder to solve than those ones from the first experiment, that is they take comparably more time. Similarly, the instances with large number of items, larger capacity, and strong correlation between profits and weights are now hard for the DP as well. Oppositely to the results of the previous experiment, the FPTAS performs much better when dealing with such instances in the case when $\epsilon$ is relaxed. For example, its performance is improved by 95% for the instance *m-s-corr_10* with 1,000 items when $\epsilon$ is raised from 0.0001 to 0.75 while the approximation rate remains at 100%.

# 5   Conclusion

Multi-component combinatorial optimization problems play an important role in many real-world applications. We have examined the non-linear Packing While Traveling Problem which results from the interactions in the Traveling Thief Problem. We designed a dynamic programming algorithm that solves the problem in pseudo-polynomial time. Furthermore, we have shown that the original objective of the problem is hard to approximate and have given an FPTAS for optimizing the amount that can be gained over the smallest possible travel cost. It should be noted that the FPTAS applies to a wider range of problems as our proof only assumed that the travel cost per unit distance in dependence of the weight is increasing and convex. Our experimental results on different types of knapsack instances show the advantage of the dynamic program over the previous approaches based on mixed integer programming and branch-infer-and-bound concepts. Furthermore, we have demonstrated the effectiveness of the FPTAS on instances with a large weight and profit range.

**Acknowledgements.** The first, second, and fifth author were supported by Australian Research Council grants DP130104395 and DP140103400. The third author is supported by the Einstein Foundation Berlin in the framework of MATHEON.

# References

1. Bonyadi, M., Michalewicz, Z., Barone, L.: The travelling thief problem: the first step in the transition from theoretical problems to realistic problems. In: 2013 IEEE Congress on Evolutionary Computation (CEC), pp. 1037–1044, June 2013. https://doi.org/10.1109/CEC.2013.6557681
2. El Yafrani, M., Ahiod, B.: Population-based vs. single-solution heuristics for the travelling thief problem. In: Proceedings of the Genetic and Evolutionary Computation Conference 2016 GECCO 2016, pp. 317–324. ACM, New York (2016). https://doi.org/10.1145/2908812.2908847
3. Faulkner, H., Polyakovskiy, S., Schultz, T., Wagner, M.: Approximate approaches to the traveling thief problem. In: Proceedings of the 2015 Annual Conference on Genetic and Evolutionary Computation GECCO 2015, pp. 385–392. ACM, New York (2015).https://doi.org/10.1145/2739480.2754716
4. GOODYEAR: Factors Affecting Truck Fuel Economy (2008). http://www.goodyeartrucktires.com/pdf/resources/publications/FactorsAffectingTruckFuelEconomy.pdf
5. Hochbaum, D.: Appromixation Algorithms for NP-hard Problems. PWS Publishing Company (1997)
6. Hoy, D., Nikolova, E.: Approximately optimal risk-averse routing policies via adaptive discretization. In: Bonet, B., Koenig, S. (eds.) Proceedings of the Twenty-Ninth AAAI Conference on Artificial Intelligence, pp. 3533–3539. AAAI Press, Austin, 25–30 January 2015
7. Lin, C., Choy, K., Ho, G., Chung, S., Lam, H.: Survey of green vehicle routing problem: past and future trends. Expert Syst. Appl. **41**(4, Part 1), 1118–1138 (2014). https://doi.org/10.1016/j.eswa.2013.07.107
8. Mei, Y., Li, X., Yao, X.: On investigation of interdependence between subproblems of the travelling thief problem. Soft Comput. **20**, 157–172 (2016). https://doi.org/10.1007/s00500-014-1487-2
9. Meyer, U., Sanders, P., Sibeyn, J. (eds.): Algorithms for Memory Hierarchies. LNCS, vol. 2625. Springer, Heidelberg (2003). https://doi.org/10.1007/3-540-36574-5
10. Pisinger, D.: Where are the hard knapsack problems? Comput. Oper. Res. **32**, 2271–2284 (2005). https://doi.org/10.1016/j.cor.2004.03.002

11. Polyakovskiy, S., Neumann, F.: The packing while traveling problem. Eur. J. Oper. Res. **258**, 424–439 (2017)
12. Polyakovskiy, S., Bonyadi, M.R., Wagner, M., Michalewicz, Z., Neumann, F.: A comprehensive benchmark set and heuristics for the traveling thief problem. In: Proceedings of the 2014 Annual Conference on Genetic and Evolutionary Computation GECCO 2014, pp. 477–484. ACM, New York (2014). https://doi.org/10.1145/2576768.2598249
13. Polyakovskiy, S., Neumann, F.: Packing while traveling: mixed integer programming for a class of nonlinear knapsack problems. In: Michel, L. (ed.) CPAIOR 2015. LNCS, vol. 9075, pp. 332–346. Springer, Cham (2015). https://doi.org/10.1007/978-3-319-18008-3_23
14. Toth, P.: Dynamic programming algorithms for the zero-one knapsack problem. Computing **25**, 29–45 (1980). https://doi.org/10.1007/BF02243880
15. Yang, G., Nikolova, E.: Approximation algorithms for route planning with nonlinear objectives. In: Schuurmans, D., Wellman, M.P. (eds.) Proceedings of the Thirtieth AAAI Conference on Artificial Intelligence, pp. 3209–3217. AAAI Press, Phoenix, 12–17 February 2016
16. Zhang, J.: A survey on streaming algorithms for massive graphs. In: Aggarwal, C., Wang, H. (eds.) Managing and Mining Graph Data. Advances in Database Systems, vol. 40, pp. 393–420. Springer, Boston (2010). https://doi.org/10.1007/978-1-4419-6045-0_13

# Multi-commodity Flow with In-Network Processing

Moses Charikar[1], Yonatan Naamad[2(✉)], Jenifer Rexford[3], and X. Kelvin Zou[4]

[1] Stanford University, Stanford, CA, USA
moses@cs.stanford.edu
[2] Amazon.com, East Palo Alto, CA, USA
ynaamad@amazon.com
[3] Princeton University, Princeton, NJ, USA
jrex@cs.princeton.edu
[4] ByteDance, Seattle, WA, USA
xuanzou1991@gmail.com

**Abstract.** Modern networks run "middleboxes" that offer services ranging from network address translation and server load balancing to firewalls, encryption, and compression. In an industry trend known as Network Functions Virtualization (NFV), these middleboxes run as virtual machines on any commodity server, and the switches steer traffic through the relevant chain of services. Network administrators must decide how many middleboxes to run, where to place them, and how to direct traffic through them, based on the traffic load and the server and network capacity. Rather than placing *specific* kinds of middleboxes on each processing node, we argue that server virtualization allows each server node to host *all* middlebox functions, and simply vary the fraction of resources devoted to each one. This extra flexibility fundamentally changes the optimization problem the network administrators must solve to a new kind of multi-commodity flow problem, where the traffic flows consume bandwidth on the links as well as processing resources on the nodes. We show that allocating resources to maximize the processed flow can be optimized exactly via a linear programming formulation, and to arbitrary accuracy via an efficient combinatorial algorithm. Our experiments with real traffic and topologies show that a joint optimization of node and link resources leads to an efficient use of bandwidth and processing capacity. We also study a class of design problems that decide where to provide node capacity to best process and route a given set of demands, and demonstrate both approximation algorithms and hardness results for these problems.

**Keywords:** Multi-commodity flow · Middleboxes ·
Network Function Virtualization · Approximation algorithms ·
Hardness of approximation

Y. Naamad—This work was done while the author was at the Department of Computer Science, Princeton University.

Y. Disser and V. S. Verykios (Eds.): ALGOCLOUD 2018, LNCS 11409, pp. 73–101, 2019.
https://doi.org/10.1007/978-3-030-19759-9_6

# 1   Introduction

In addition to delivering data efficiently, modern networks often perform services on the traffic in flight to enhance security, privacy, or performance, or provide new features. Network administrators often install "middleboxes" such as firewalls, network address translators, server load balancers, Web caches, video transcoders, and devices that compress or encrypt the traffic. In fact, many networks have as many middleboxes as underlying routers or switches [29]. Often a single connection must traverse multiple middleboxes, and different connections may go through different sequences of middleboxes. For example, while Web traffic may go through a firewall followed by a server load balancer, video traffic may simply go through a transcoder. To keep up with the traffic demands, an organization may run multiple instances of the same middlebox. Deciding how many middleboxes to run, where to place them, and how to direct traffic through them, is a major challenge facing network administrators.

Until recently, each middlebox was a dedicated appliance, consisting of both software and hardware. Administrators typically installed these appliances at critical locations that naturally see most of the traffic, such as the gateway connecting a campus or company to the rest of the Internet. A network could easily have a long chain of these appliances at one location, forcing all connections to traverse every appliance—whether they need all of the services or not. In addition, placing middleboxes only at the gateway does not serve the organization's many *internal* connections, unless the internal traffic is routed circuitously through the gateway.

Over the last few years, middleboxes have become increasingly *virtualized*, with the software service separate from the physical hardware—an industry trend called Network Functions Virtualization (NFV) [5,22]. The network can "spin up" (or down) virtual machines on any physical server, as needed. This has led to a growing interest in good algorithms for optimizing the (i) *allocation* of middleboxes over a pool of server resources, (ii) *steering* of traffic through a suitable sequence of middleboxes based on a high-level policy, and (iii) *routing* of the traffic between the servers over efficient network paths [1,16,17,19,24].

Rather than solving these three optimization problems separately, we introduce—and solve—a joint optimization problem. Since server resources are fungible, we argue that each processing node could subdivide its resources *arbitrarily* across any of the middlebox functions, as needed. That is, the *allocation* problem is more naturally a question of what fraction of each node's computational (or memory) resources to allocate to each middlebox function. Similarly, each connection can have its middlebox processing performed on any node, or set of nodes, that have sufficient resources. That is, the *steering* problem is more naturally a question of deciding which nodes should devote a share of their processing resources to a particular portion of the traffic. Hence, the joint optimization problem devolves to a new kind of *routing* problem, where we compute paths based on both the bandwidth and processing requirements of the traffic between each source-sink pair. That is, each flow from a source to a sink must be allocated both (i) a certain amount of bandwidth on *every* link in its path and (ii) a *total* amount of computational across all of the nodes on its path.

In our *flow with in-network processing* problem, we have a flow demand with multiple sources and multiple sinks, and each flow requires a certain amount of processing. The required processing is proportional to the flow size and, without loss of generality, we assume one unit of flow requires one unit of processing. Each flow from a source to a sink is an aggregate flow of many connections, so the routing and processing for a flow are both divisible. In this model there are two types of constraints: edge capacity constraints and vertex capacity constraints, which represent link bandwidth and node processing capacity, respectively. A feasible flow pattern satisfies three conditions: (i) for each edge, the sum over all flows on that edge is bounded by that edge's capacity, (ii) for each node, the sum over all flows of in-network processing done at that node is bounded by the vertex capacity, and (iii) each flow must be allocated a total amount of node processing power equal to its size.

Although ignoring vertex capacity constraints reduces our class of problems to those of the standard multi-commodity flow variety, the introduction of these constraints yields a new class of problems that has not been studied before. This paper provides a systematic approach to this new class of network problems, applicable to both directed and undirected graphs.

In Sect. 2, we introduce the PROCESSED FLOW ROUTING class of problems, in which we discuss how to optimize processed flow routed in a fixed network. Next, we present two linear programming-based algorithms to find a maximum feasible multi-commodity flow with the additional processing constraints. We show that, like standard multi-commodity flow, the program can be written in two different equivalent ways: either with an exponentially-sized walk-based LP or with a polynomially-sized edge-based LP. The proof of equivalence of these two LPs requires a more careful argument than that for standard MCF. As an aside, we argue that this pair of LPs can also be adapted to optimize several other objective functions, such as those minimizing congestion. In Sect. 4, we briefly describe an experimental evaluation of this linear programming approach.

In Sect. 5, we design an efficient multiplicative weight update (MWU) algorithm that finds approximately optimal solutions to our walk-based linear program far more quickly than one could with the edge-based program paired with an off-the-shelf LP-solver.

In Sect. 6, we consider the MIDDLEBOX NODE PURCHASE class of problems, in which the goal is to optimally purchase processing power at various middleboxes. Prices for placing processing at the various nodes is given as part of the input, and may differ substantially from one location to the next. This class of problems comes has two natural variants:

1. MIN MIDDLEBOX NODE PURCHASE: given a set of flow demands, minimize cost while purchasing enough middlebox processing capacity so that all flow demands are simultaneously satisfiable (that is, jointly routable and steerable).

2. BUDGETED MIDDLEBOX NODE PURCHASE: given a set of flow demands and a budget of $k$ dollars, spend at most \$$k$ on purchasing middlebox processing capacity while maximizing the fraction of the given demand that is simultaneously satisfiable.

Linear programs for both of these problems can be found in Sects. 6.2 and 6.4. For MIN MIDDLEBOX NODE PURCHASE, we show an $O(\log(n)/\delta^2)$ approximation for node costs and an associated multi-commodity flow that satisfies $(1 - \delta)$ fraction of the demands and satisfies all edge capacities, where $n$ is the number of nodes. We show that in the directed case, the problem is hard to approximate better than a logarithmic factor, even if the demand requirements are relaxed. Additionally, we show that the undirected case is at least as hard to approximate as VERTEX COVER.

We also prove approximation and hardness results for BUDGETED MIDDLEBOX NODE PURCHASE. Although it's tempting to conjecture that the problem is an instance of BUDGETED SUBMODULAR MAXIMIZATION, one can construct instances on both directed and undirected graphs where the amount of routable processed flow is *not* submodular in the set of purchased nodes, so black-box submodular maximization techniques cannot be used here. We show an $\Omega(1/\log(n))$ approximation for both problems, as well as a constant factor approximation algorithm for undirected instances with a single source-sink pair. For the directed case, we show approximation hardness of $1 - 1/e$, and constant factor hardness for the undirected problem. Our results are summarized in the following Table 1

**Table 1.** Network design results

|  |  | Directed | Undirected |
|---|---|---|---|
| Budgeted | Approximation hardness | $\Omega(1/\log n)$ | $.078^{\text{b}}$ |
|  |  | $1 - 1/e$ | $.999$ |
| Minimization | Approximation hardness | $O(\log n)^{\text{a}}$ | $O(\log n)^{\text{a}}$ |
|  |  | $O(\log n)$ | $2 - \epsilon$ |

[a]All demands are satisfied only up to an $(1 - \epsilon)$ fraction.
[b]Assuming 1 source-sink pair. For multiple pairs, we adapt the $\Omega(1/\log n)$-approximation digraph algorithm.

# 2    Flow Routing with In-Network Processing

## 2.1    Processed Flow Routing Problem

Network Function Virtualization (NFV) allows each node to function as a general-purpose server that can run any in-network processing task, such as transcoding, compression, and encryption. Such servers can reside anywhere in the network, from the leaf nodes (as in the case of traditional servers) to intermediary nodes (such as top-of-rack and spine switches).

Therefore, in our model, we treat all in-network processing as *homogeneous*, meaning that every node with a sufficient quantity of available computational

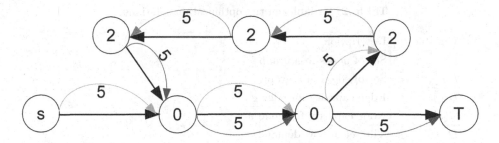

**Fig. 1.** The edge capacity is 10 for all edges and the node capacities are denoted in each node. Here, we can send maximum flow size 5, by routing it along the red arcs, have it processed at the nodes at the top, and then sent to $T$ along the blue arcs. The capacity of the bottom middle edge forms the bottleneck here, as all flow must pass through it twice before reaching $T$. (Color figure online)

resources can be adapted to accomplish any processing task. In practice, this can be accomplished simply by spinning up a new virtual machine for that specific task as needed. We assume that all flows are both aggregate and sufficiently large that they can be treated as continuous quantities (and thus can be arbitrarily subdivided), and that the processing capacity of a given node can also be fractionally divided among a number of different flows.

Each flow is initially generated at a source fully unprocessed. By the time it reaches its destination, it needs to go through and get processed by one or more intermediate processing nodes with available computational resources. We assume that each unit of flow requires one unit of processing, meaning that for any given flow $f$, the total processing workload done on $f$ by vertices along $f$'s flow path should equal the size $f$.

This problem can be modeled mathematically as follows. We are given an (un)directed graph $G = (V, E)$ along with edge capacities $B : E \rightarrow \mathbb{R}^+$, vertex capacities $C : V \rightarrow [0, \infty)$, and a collection of flows of varying commodities $D = \{(s_1, t_1, k_1), (s_2, t_2, k_2), \cdots\} \subseteq V \times V \times \mathbb{R}^+$. While the edge capacities are used in the same way as in a standard multi-commodity flow problem, we also require that each unit of flow undergo a total of one unit of processing at intermediate vertices. In particular, while edge capacities limit the *total* amount of flow that may pass through an edge, vertex capacities only bottleneck the amount of processing that may be done at a given vertex, regardless of the total amount of flow that uses the vertex as an intermediate node. The goal is then either to route as much flow as possible, or to satisfy all flow demand subject to a congestion-minimization objective function. For concreteness, this paper focuses on maximizing the total amount of flow we can send between the source-destination pairs while satisfying edge and node capacity constraints. In practice we can also extend our results to other objective functions such as minimizing the weighted sum of congestion at edges and nodes. For ease of exposition, we focus our attention almost entirely on directed instances. The results for undirected graphs follow analogously (Table 2).

**Table 2.** Variables in the optimization solutions

| Variable | Description |
|---|---|
| $V$ | Set of nodes in a graph |
| $E$ | Set of edges in a graph |
| $B(e)$ | Edge capacity for edge $e$ |
| $C(v)$ | Node capacity for node $v$ |
| $D$ | The set of flow demands |
| $\delta^+(v)$ | The edges leaving vertex $v$ |
| $\delta^-(v)$ | The edges entering vertex $v$ |
| $P$ | The set of 2-walks from sources to destinations |
| $p_{i,\pi}^v$ | 2-walk-based; the amount of flow $i$ from $s_i$ to $t_i$ exactly using 2-walk $\pi$ and processed at $v$ |
| $f_i(e)$ | Edge-based; the amount of flow $i$ that traverses $e$ on its way from $s_i$ to $t_i$ |
| $w_i(e)$ | Edge-based; the amount of unprocessed flow $i$ that traverses $e$ on its way from $s_i$ to $t_i$ |
| $p_i(v)$ | Edge-based; the amount of processing done at node $v$ for the $i$th flow |

## 2.2 A 2-Walk-Based Solution

We now describe a *2-walk-based* formulation of the problem. A *2-walk* from $s$ to $t$ is a route between $s$ and $t$ that visits each vertex (and thus each edge) at most *twice*.

The approach we take is analogous to *path-based* solutions for the traditional multicommodity flow (MCF) problem, with the key difference that, unlike paths, our 2-walks may visit vertices and edges more than once. Additionally, a 2-walk may traverse the same undirected edge in both directions.

To express the *2-walk-based* linear program, we introduce one variable $p_{i,\pi}^v$ for each {2-Walk}–vertex–demand triplet, representing the total amount of flow from $s_i, t_i$ exactly utilizing walk $\pi$ and processed at $v$. Note here the set $P$ of 2-walks is an enumeration of all possible 2-walks in the graph, which can be *exponential* in size. The LP is then formulated as follows:

MAXIMIZE
$$\sum_{i=1}^{|D|} \sum_{\pi \in P} \sum_{v \in \pi} p_{i,\pi}^v$$

SUBJECT TO

$$p_{i,\pi} = \sum_{v \in \pi} p_{i,\pi}^v \qquad \forall i \in [|D|], \forall \pi \in P$$

$$\sum_{i=1}^{|D|} \sum_{\substack{\pi \in P \\ \pi \ni e}} p_{i,\pi} \leq B(e) \qquad \forall e \in E$$

$$\sum_{i=1}^{|D|} \sum_{\pi \in P} p_{i,\pi}^v \leq C(v) \qquad \forall v \in V$$

$$\sum_{\pi \in P} \sum_{v \in \pi} p_{i,\pi}^v \leq k_i \qquad\qquad \forall i \in [|D|]$$

$$p_{i,\pi}^v \geq 0 \qquad\qquad \forall i \in [|D|], \forall \pi \in P, \forall v \in V$$

While the first constraint enforces that all flows are fully processed, the second and third constraints ensure that no edge or vertex is over-saturated.

# 3   An Edge-Based Polynomially-Sized LP

Although the 2-walk-based solution exactly solves our MCF with in-network processing problem, the LP may be exponentially sized and thus even writing it down (let alone solving it) leaves us with an exponential worst-case running time. In Sect. 3.1, we present a polynomially-sized (and thus polytime-solvable) edge-based linear program for this problem. We then follow this up by a proof of correctness in Sect. 3.2.

## 3.1   The Edge-Based Solution

A standard technique for solving the traditional MCF problem relies on constructing a polynomially-sized *edge-based* LP whose set of feasible solutions equals that of an exponentially-sized path-based LP. Analogously, we establish a polynomial-sized edge-based LP corresponding to the *2-walk-based* LP introduced previously.

$$\text{MAXIMIZE} \qquad\qquad\qquad\qquad \sum_{i=1}^{|D|} \sum_{e \in \delta^+(s_i)} f_i(e) \qquad (2a)$$

$$\text{SUBJECT TO} \qquad\qquad\qquad\qquad\qquad\qquad\qquad (2b)$$

$$\sum_{e \in \delta^-(v)} f_i(e) = \sum_{e \in \delta^+(v)} f_i(e) \qquad \forall i \in [|D|], \forall v \in V \setminus \{s_i, t_i\} \qquad (2c)$$

$$p_i(v) = \sum_{e \in \delta^-(v)} w_i(e) - \sum_{e \in \delta^+(v)} w_i(e) \qquad \forall i \in [|D|], \forall v \in V \setminus \{s_i\} \qquad (2d)$$

$$\sum_{i=1}^{|D|} f_i(e) \leq B(e) \qquad\qquad\qquad \forall e \in E \qquad (2e)$$

$$\sum_{i=1}^{|D|} p_i(v) \leq C(v) \qquad\qquad\qquad \forall v \in V \qquad (2f)$$

$$\sum_{e \in \delta^+(s_i)} f_i(e) \leq k_i \qquad\qquad\qquad \forall i \in [D] \qquad (2g)$$

$$w_i(e) \leq f_i(e) \qquad\qquad\qquad \forall i \in [D], \forall e \in E \qquad (2h)$$

$$w_i(e) = f_i(e) \qquad\qquad\qquad \forall i \in [D], \forall e \in \delta^+(s_i) \qquad (2i)$$

$$w_i(e) = 0 \qquad\qquad\qquad \forall i \in [D], \forall e \in \delta^-(t_i) \qquad (2j)$$

$$w_i(e), p_i(v) \geq 0 \qquad\qquad\qquad \forall i \in [D], \forall e \in E \qquad (2k)$$

The LP constraints can be interpreted as follows. Constraint (2c) is a flow conservation constraint: at any non-terminal node of flow $i$, the amount of flow $i$ that enters the node equals the amount that leaves it. Constraint (2d) is a *processing conservation* constraint, ensuring that the total amount of flow (processed or unprocessed) going through a node remains unchanged, although the quantity of each might change if the node processes any of the flow. Constraints (2e) and (2f) ensure that we don't exceed edge and node capacities. Constraint (2g) ensures that we don't route more flow than is requested between any demand pair. Constraint (2h) ensures that the amount of work yet to be done on a flow does not exceed the size of the flow itself, while (2i) and (2j) ensure that all flows leave the sources unprocessed and arrive to the destinations fully processed.

## 3.2   Proof of Equivalence to the 2-Walk-Based LP

While the construction of the edge-based LP is not particularly difficult, it is not obvious that the edge-based solution actually solves the problem in question. We need to prove the *correctness* of the edge-based LP. A priori, solutions to the edge-based LP here may not be decomposable to a valid routing pattern at all. We provide an efficient algorithm converting feasible solutions to the edge-based LP into corresponding solutions to the 2-walk-based program, proving both that the edge-based LP is correct and that the actual flow paths can be recovered in polynomial time as well. We summarize this result in the following theorem.

**Theorem 1.** *The edge-based formulation provides a polynomial-sized linear program solving the Maximum Processed Flow problem. Further, the full routing pattern can be extracted from the LP solution by decomposing it into its composing $s_i, t_i$ 2-walks in $O(|V| \cdot |E| \cdot |D| \cdot \log|V|)$ time.*

Notably, as the reduction maps the set of feasible solutions to the edge-based LP to equivalent feasible solutions of the 2-walk-based LP, the same technique can also be used to show the equivalence of the two corresponding programs when the objective function is changed to optimize some other linear quantity, such as the amount of congestion.

We now describe the algorithm in more detail and prove its correctness. The first part of the proof involves showing how to construct a solution to the flow-based LP when there is exactly one $s_i, t_i$ pair. Extracting the corresponding flow paths and iterating this procedure for each demand pair eventually extracts all $s_i, t_i$ flows, giving us a solution to the multicommodity problem.

The flow extraction argument proceeds in two steps. First, we simplify the solution by removing extraneous loops that do not affect the optimal solution. Next, we show that the existence of any residual flow in the graph (i.e., the existence of some strictly positive $f_i(e)$) implies that there exists at least one valid 2-walk we can efficiently extract while maintaining feasibility of all constraints for the updated residual graph. As we show, a linear number of extractions suffices to remove all flow from the solution. We provide a complete algorithm in Algorithm 1.

## Algorithm 1. 2-Walk Decomposition

**Data:** $G'(V, E)$, $w(e)$, $f(e)$ for $\forall e \in E$ and $p(v)$ for $\forall v \in V$
**Result:** $f(\pi)$, $p(\pi, v)$ with $v \in \pi$
**Algorithm** TwoWalkConstruction$(s, t, v)$

> //*Construct 2-walk from* $s \to v$ *and* $v \to t$
> From $v$, run a backward traversal, each time picking an incoming edge $e$
> maximizing $\rho(e) = w(e)/f(e)$
> From $v$, run a forward traversal, each time picking an outgoing edge minimizing
> $\rho(e) = w(e)/f(e)$.
> **return** $\pi$

**Algorithm** FlowPlacement$(s, t)$

> **while** *there exists a* $v$ *with* $p(v) > 0$ **do**
> > $\pi \leftarrow$ TwoWalkConstruction$(s, t, v)$
> > $f' \leftarrow \min_{e \in \pi, e \text{ precedes } v} f^1(e)$
> > $f'' \leftarrow \min_{e \in \pi, e \text{ succeeds } v} f^2(e)$
> > $p_\pi^v \leftarrow \min\{f', f'', p(v)\}$
> > **for** $u \in \pi$ *and* $u \neq v$ **do**
> > > $p_\pi^u = 0$
> >
> > **end**
> > $C(v) \leftarrow C(v) - p_\pi^v$
> > $p(v) \leftarrow p(v) - p_\pi^v$
> > **for** $e \in \pi$ **do**
> > > $f(e) \leftarrow f(e) - p_\pi^v$
> > > $B(e) \leftarrow B(e) - p_\pi^v$
> >
> > **end**
>
> **end**

**Removing Extraneous Loops.** Suppose we are given a nonempty solution to the *edge-based* LP for an instance with graph $G(V, E)$. We focus on some (arbitrarily chosen) commodity $i$ with positive flow in this solution, and drop subscripts to let $f(e)$, $w(e)$, and $p(v)$ denote $f_i(e)$, $w_i(e)$, and $p_i(v)$, respectively. To assist with our exposition, we restrict our attention to the subgraph $G'$ which excludes all edges for which $f(e) = 0$. For each edge $e$ in this subgraph, we also associate two new variables, $f^1(e)$ and $f^2(e)$ denoting the amount of unprocessed and processed flow passing through this edge, respectively. Thus, by definition, $f^1(e) = w(e)$ and $f^2(e) = f(e) - w(e)$.

As in solutions to the edge-based linear program for the standard multicommodity flow problem, solutions to our edge-based LP may introduce closed loops (that is, directed cycles along which a positive amount of flow is routed). In traditional MCF, such loops are easily shown to be *non-essential*, and can be easily removed from a feasible solution without affecting its correctness. As illustrated in Fig. 1, such loops may actually be critical in solutions to our variant, and handling such cases takes additional care. Thus, instead of arguing that cycles can be removed (so that the flows form a set of paths), we show how to ensure that no vertex may be visited more than twice (and thus the flows form a set of

2-walks). In particular, we show how to cancel out all cycles along which each edge contains $f^1$ flow, as well as all cycles along which each edge carries $f^2$ flow.

**Lemma 1.** *Any closed loop for which every edge contains $f^1$ (resp. $f^2$) flow can be removed without affecting the total $(s, t)$ flow.*

*Proof.* This argument proceeds similarly to the flow cancellation arguments in the traditional MCF setting. For any loop $l$ containing a positive amount of $f_1$ flow, reducing both $f(e)$ and $w(e)$ on the constituent edges by $\min_{e' \in l} f^1(e')$ ensures that all constraints in the LP remain satisfied. For loops containing a positive amount of $f_2$ flow, similarly reducing just $f(e)$ suffices.

**Extracting 2-Walks.** Suppose extraneous loops have been removed using the process described in Lemma 1. Define $\rho_e = \frac{w(e)}{f(e)} = \frac{f^1(e)}{f^1(e) + f^2(e)}$. By Lemma 1, every cycle with a positive $f(e)$ on each edge contains at least one edge with $\rho = 1$ and another with $\rho = 0$. We now repeat the following until all flow is removed from the graph. Select a vertex $v$ that is allocated processing (i.e., $p(v) > 0$), and run a backwards traversal from $v$, at each step selecting the incoming edge with the largest fraction of unprocessed flow (i.e., maximizing $\rho(e)$) until we reach $s$. Similarly, run a forward traversal from $v$ to $t$ along edges minimizing $\rho$. This route will be our "flow-2-walk". The amount of flow routable along this flow-2-walk is the minimum of three quantities: (1) the smallest amount of unprocessed flow sent on each edge of the $s \leadsto v$ path, (2) the smallest amount of processed flow sent along each edge of the $v \leadsto t$ path, and (3) the amount of processing still to be done at $v$ (i.e., $p(v)$). We then extract this flow-2-walk from the solution by decreasing each LP variable accordingly. Complete pseudo-code for this algorithm is given in Algorithm 1.

**Lemma 2 (2-Walk Extraction).** *Algorithm 1 can always generate a 2-walk with non-zero flow from source to sink if there exists any $v$ where $p(v) > 0$. Further, the number of iterations needed of Algorithm 1 is bounded by $O(|E|)$, each of which can be made to take $O(|V| \log |V|)$ time. Thus, the total running time is $O(|E| \cdot |V| \log |V|)$.*

*Proof.* The removal of extraneous cycles guarantees that no 2-walk can visit the same vertex more than twice. Now suppose that a vertex $v$ has $p(v) > 0$. By constraint (2d), the $f^1$ flow on some incoming edge and the $f^2$ flow on some outgoing edge must both be positive. By a combination of constraints (2c), (2d), and (2j), the reverse traversal from $v$ to $s$ must succeed: it cannot get "stuck" at a vertex $u$ with no in-edge with positive $f^1$ flow. Similarly, the forward traversal from $v$ to $t$ must find a path with positive $f^2$ on each edge. Subtracting the minimum of all of the reverse path's observed $f^1$ values, all of the forward path's observed $f^2$ values, and $p(v)$ from each of those variables ensures that all variables remain nonnegative. Further, as this operation is monotone and it decreases one of the variables to 0, repeating this must remove all flow from the graph in at most $|V| + 2|E| = O(|E|)$ iterations. By initially constructing a priority queue for each vertex on the $f^1$, $f^2$, and $\rho$ values of its neighboring edges and updating them accordingly, the forward and backward traversals can be found in $|V| \log |V|$ time, each.

We can generalize the above approach to the multicommodity problem by treating each of the commodities independently. Namely, sequentially applying the above algorithm to remove flow 2-walks for each of the $|D|$ demand pair gives us a solution to the multicommodity problem without violating any of the LP constraints. Thus, we get the $O(|V| \cdot |E| \cdot |D| \cdot \log |V|)$ running time promised in the statement of Theorem 1.

## 4 Evaluations

We ran several experiments to address (i) how well the LP fares against "naive" algorithms, and (ii) the in-practice running time for an edge-based LP solution.

**Throughput Improvement.** To determine how well the LP fares against simple approaches, we compare it to a "naive" algorithm that first routes flow without vertex capacities in mind, and then processes as much flow as possible on the flow paths it initially routed. This is a variant of the *path-selection* approach used in [15]. While there are simple examples where the naive algorithm performs extremely poorly in theory, we seek to study the performance in practice.

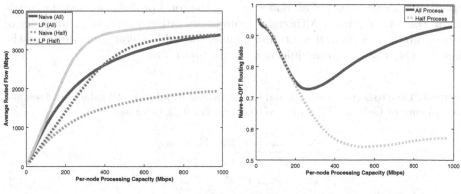

(a) Amount of flow that the two algorithms could process given various node processing capacities.

(b) The ratio of the demanded flow processable by the naive algorithm to that processable by the LP, plotted at various processing capacities.

**Fig. 2.** Experimental results showing how much of the demand both the naive and the (optimal) LP-based algorithm could successfully route and process given the Abilene traffic matrices.

We ran both algorithms on 150 randomly sampled traffic matrices provided by the TOTEM project [31] for the Abilene network in 2004. As these datasets don't include vertex processing capacities, we compared the two algorithms on a wide range of values, with processing capacities assigned according to one of two distributions: either they all have the same capacity (the *all* case) or exactly half of them have the prescribed capacity and the other half have zero (the *half* case). The results are diagrammed in Fig. 2.

Experimental analysis shows that while the LP and the "naive" algorithm fare similarly when the network is low on processing capacity and thus node-throttled, or, in the *all* case, high on node capacity and thus bottlenecked by the link capacities and the demand itself, the LP has a distinct advantage in between the two extremes when either resource could become the bottleneck when the flows are not routed efficiently. Additionally, the experiments show that the naive algorithm suffers when processing is not uniformly distributed among the nodes even in the high-capacity case, as many of the initial flow paths might go entirely through nodes without any processing capacity and thus fail to get processed. Our experiments show that using the exact algorithm gives an improvement of up to 30% over the naive approach if processing power is available at all nodes, and up to 80% if the processing power is only placed at half of the nodes.

**Runtime Analysis.** Although the edge-based LP provides a polynomial running-time guarantee, it may still be too slow in practice on large graphs. To study the empirical performance of linear programming, we also run the LP solver over a number of topologies acquired from SNDLib [23]. For each of the topologies in Table 3, enough processing capacity was evenly distributed among a random sample of half of all nodes so that the total processing capacity equals half of the total demand. Although the implementation, hardware, and choice of solver were not optimized for running time, the table below indicates that the time to solve the LP grows quickly with the input size.

**Table 3.** Time to solve the edge based LP for various topologies. All values are averaged over 15 runs of CoinLP [13] on a 3.3 GHz Intel i5 2500k processor.

| Network | $|V|$ | $|E|$ | Time (sec) |
|---|---|---|---|
| abilene | 12 | 15 | 1.91 |
| dfn-bwin | 10 | 45 | 3.08 |
| atlanta | 15 | 22 | 5.28 |
| dfn-gwin | 11 | 47 | 13.91 |
| geant | 22 | 36 | 23.69 |
| france | 25 | 45 | 44.38 |
| india35 | 35 | 57 | 105.89 |

The cost of solving this LP even on small topologies justifies the use of the faster multiplicative weight algorithm instead. The MWU algorithm has a running time of roughly $\tilde{O}(|D|*|E|^2/\epsilon^2)$, which on sparse graphs is roughly equal to just the number of variables in the edge-based LP (as opposed to the time needed to actually solve it). While the algorithm is only approximately optimal, choosing an appropriate value of $\epsilon$ (say, $\epsilon = 0.1$) can grant a better running time while still significantly outperforming the naive algorithm.

# 5  Multiplicative Weights Based Approximation Algorithm

We first briefly overview the MWU method in Sect. 5.1. Next, we describe how to apply the MWU method to our model including processing vertices. The proof of correctness is given in Sect. 5.3;

## 5.1  Multiplicative Weight Update for Traditional MCF

In the traditional multiplicative weights algorithm for multicommodity flow, there an "expert" is assigned to each edge, each of which is initially assigned a sufficiently small weight. The algorithm then iteratively finds $s_i, t_i$ walks minimizing the sum of weighted utilization of their edges and adds together scaled down versions of these paths to eventually construct a solution. When a path is chosen, all experts corresponding to edges along the path have their weight increased by a multiplicative factor, making it less likely that we repeat our selection of the edges. This process is repeated until some expert's weight surpasses the value 1, corresponding to a fully utilized edge. When this happens, all paths are scaled down by the weight of the largest expert to ensure that no capacities are exceeded. One then shows that the final result is within a $(1 - \epsilon)$ factor of the maximum multicommodity flow.

## 5.2  Formulation and Analysis

Although we derive the same $(1 - \epsilon)$ approximation factor for our problem, the analysis of our multiplicative weights algorithm is quite different from that of traditional multicommodity flow. Intuitively, this is because vertex capacities are inherently very different from edge capacities: while a flow 2-walk reduces the remaining capacity on *all* edges it traverses, it only reduces the capacity for *one* of its vertices. Thus, we set up a different update condition, as well as a different method for picking the best flow 2-walks for each round.

**Setup.** For each edge $e$, we have a constraint $\sum_\pi p_\pi \leq B(e)$, where $p_\pi$ is the amount of flow sent on 2-walk $\pi$. For each vertex, the corresponding constraint is $\sum_\pi p_\pi^v \leq C(v)$, where $p_\pi^v$ is the amount of flow on 2-walk $\pi$ that is processed at $v$. For each of these two sets of constraints, we associate one expert, (which we call $\hat{e}$ and $\hat{v}$), whose weights are denoted by $w_{\hat{e}}$ and $q_{\hat{v}}$, respectively.

Consider a feasible solution to the 2-walk-based LP. The feasible solution consists of variables of the form $p_\pi$ and $p_\pi^v$. In this section, we abuse notation and let the variable $p$ denote a feasible solution to the LP, at which point $p_\pi$ and $p_\pi^v$ become bound variables for each $\pi$ and $v$ (that is, $p$ can be thought of as a dictionary containing the aforementioned set of variables). Further, define $A(p)$ as the objective function value of $p$, i.e. $A(p) = \sum_{v \in V} \sum_{\pi \in P} p_\pi^v$.

For an expert $\hat{e}$ and feasible solution $p$, define the gain $M(\hat{e}, p)$ by $M(\hat{e}, p) = \frac{1}{B(e)} \sum_{\pi \ni e} p_\pi$. This can be thought of as the fraction of $e$'s capacity actually utilized by the feasible solution. For each expert $\hat{v}$, we define the gain $M(\hat{v}, p)$ by $M(\hat{v}, p) = \frac{1}{C(v)} \sum_{\pi \ni v} p_\pi^v$, which corresponds to the fractional utilization of $v$'s processing capacity.

Let $\mathcal{D}$ be the probability distribution over experts in which the probability of choosing a given expert is proportional to its weight. For a fixed $p$, the expected gain of a random variable sampled from $\mathcal{D}$ is

$$M(\mathcal{D}, p) = \frac{\sum_e w_{\hat{e}} M(\hat{e}, p) + \sum_v q_{\hat{v}} M(\hat{v}, p)}{\sum_e w_{\hat{e}} + \sum_v q_{\hat{v}}}$$

We first make two observations:

**Observation 1:** For any feasible solution $p$, $0 \leq M(\mathcal{D}, p) \leq 1$. This is because $M(\hat{e}, p) \leq 1$ and $M(\hat{v}, p) \leq 1$ for all $e$ and $v$.

**Observation 2:** For any feasible solution $p$ and weights $w, q$, if $\pi^* = \operatorname{argmin}_\pi \left( \sum_{e \in \pi} w_{\hat{e}}/B(e) + \min_{\hat{v} \in \pi} q_{\hat{v}}/C(v) \right)$, then

$$M(\mathcal{D}, p) \geq \frac{A(p) \left( \sum_{\hat{e} \in \pi^*} w_{\hat{e}}/B(e) + \min_{\hat{v} \in \pi^*} q_{\hat{v}}/C(v) \right)}{\sum_e w_{\hat{e}} + \sum_v q_{\hat{v}}}$$

This is due to the fact that:

$$\begin{aligned}
M(\mathcal{D}, p) &= \frac{\sum_e w_{\hat{e}} M(\hat{e}, p) + \sum_v q_{\hat{v}} M(\hat{v}, p)}{\sum_e w_{\hat{e}} + \sum_v q_{\hat{v}}} \\
&= \frac{\sum_\pi \left( p_\pi (\sum_e w_{\hat{e}}/B(e)) + \sum_{v \in \pi} p_\pi^v q_{\hat{v}}/C(v) \right)}{\sum_e w_{\hat{e}} + \sum_v q_{\hat{v}}} \\
&\geq \frac{\sum_\pi \left( p_\pi (\sum_e w_{\hat{e}}/B(e) + \min_{v \in \pi} q_{\hat{v}}/C(v) \right)}{\sum_e w_{\hat{e}} + \sum_v q_{\hat{v}}} \\
&\geq \frac{\sum_\pi p_\pi \min_\pi (\sum_{e \in \pi} w_{\hat{e}}/B(e) + \min_{v \in \pi} q_{\hat{v}}/C(v))}{\sum_e w_{\hat{e}} + \sum_v q_{\hat{v}}} \\
&\geq \frac{A(p)(\sum_{e \in \pi^*} w_{\hat{e}}/B(e) + \min_{v \in \pi^*} q_{\hat{v}}/C(v))}{\sum_e w_{\hat{e}} + \sum_v q_{\hat{v}}}
\end{aligned}$$

Where $\pi^*$ is the path minimizing the argmin in the statement of the observation. Thus, in each round, we aim to find the $\pi^*$ minimizing this value. Conditioned on us being able to do so, the rest of the MWU algorithm proceeds as follows:

1. We initialize all expert weights $\{w_{\hat{e}}\}$ and $\{q_{\hat{v}}\}$ to $1/\delta$, where $\delta = (1+\epsilon)((1+\epsilon) \cdot |E|)^{-1/\epsilon}$. This choice of $\delta$ will be justified in the analysis of Sect. 5.3.

2. At each step $t$, given weights $w_e^t$ and $q_v^t$ on the experts, we pick the flow-2-walk $p^t$ minimizing the quantity $\sum_{e \in \pi} \frac{w_{\hat{e}}}{B(e)} + \min_{v \in \pi} \frac{q_{\hat{v}}}{C(v)}$. An efficient algorithm for finding such a 2-walk is given in Sect. 5.2.
3. Given the 2-walk $p^t$ chosen in the previous step, we treat this as a feasible solution to the instance, giving expert $\hat{j}$ a gain of $M(\hat{j}, p^t)$. Consequently, the weight $w_{\hat{e}}$ or $q_{\hat{v}}$ of each expert $j$ is increased by a multiplicative factor of $M(\hat{j}, p^t)$.
4. The algorithm stops when one of the weights $w_{\hat{e}}$ or $q_{\hat{v}}$ is larger than 1. Once the algorithm terminates, we scale down the flow $p^t$ computed at each round by a factor of $\log_{1+\epsilon} \frac{1+\epsilon}{\delta} = 1 - \frac{\ln \delta}{\ln 1+\epsilon}$, and return the set of all flow-2-walks $p^t$.

**Computing the Optimal Path.** To compute the 2-walk $\pi^t$ with minimum cost, we use a dynamic programming algorithm reminiscent of Dijkstra's shortest path algorithm. Given a graph $G(V, E)$, with weights $w(e)$ on edges, weights $n(v)$ on nodes, and some source-sink pair $s, t$, we are interested in computing the following quantity

$$opt(s,t) := \underset{\pi=(s,\cdots,t), v \in \pi}{\text{argmin}} \; cost(\pi, v) \tag{3}$$

where $cost(\pi, v)$ is defined as

$$cost(\pi, v) := \left( \sum_{e \in \pi} w(e) + n(v) \right)$$

We compute $opt(s,t)$ in two stages. First, for every $v$, we upper bound the value of $opt(s, v)$ by $n(v)$ plus the shortest distance from $s$ to $v$. Afterwards, we use dynamic programming to iteratively decrease these upper bounds. Full details are given in Algorithm 2.

---

**Algorithm 2.** Optimal 2-Walk Algorithm

---

**Require:** Graph $G = (V, E)$ with edge weights $w(e)$, node weights $n(v)$, and a designated source $s$.
    **return** $r(v) = opt(s, v)$ for every $v \in V$.
    Use Dijsktra's algorithm to compute the shortest path $d(v)$ between $s$ and $v$.
    Initialize $r(v) \leftarrow d(v) + n(v)$ for all $v \in V$. $S \leftarrow \{s\}$.
    **while** $S \neq V$ **do**
        Let $u^* = \text{argmin}_{v \in V \setminus S} r(v)$. Add $u^*$ to $S$.
        For all neighbors $z$ of $u^*$ that are not already in $S$, let $r(z) \leftarrow \min\{r(u^*) + w(u, z), r(z)\}$
    **end while**

---

**Update.** Suppose now that the 2-walk $\pi$ with smallest cost has been computed. One of two things may bottleneck the amount of processed flow that can be sent along $\pi$: either the edge capacity of some edge $e$, or the processing capacity of some vertex $v$. We consider the two cases separately

If the bottleneck is edge-based, i.e. $\sum_{v \in \pi^t} C(v) \geq \min_{e \in \pi^t} B(e)$, then let $e^t = \operatorname{argmin}_{e \in \pi^t} B(e)$, and let the chosen flow 2-walk $p^t$ be the one satisfying

$$
p_\pi^{t,v} = \begin{cases} \dfrac{C(v)}{\sum_{v \in \pi^t} C(v)} \cdot B_{e^t} & \text{if } \pi = \pi^t, v \in \pi^t \\ 0 & \text{otherwise} \end{cases}
$$

On the other hand, if $\sum_{v \in \pi^t} C(v) < \min_{e \in \pi^t} B(e)$, select $p^t$ to satisfy

$$
p_\pi^{t,v} = \begin{cases} C(v) & \text{if } \pi = \pi^t, v \in \pi^t \\ 0 & \text{otherwise} \end{cases}
$$

### 5.3   Proof of the $(1 - \epsilon)$ Approximation

Let $T$ be the number of rounds taken until we hit the stopping criterion, and let $\bar{p} = \sum_{t=1}^{T} p^t$ be the total amount of flow selected after $T$ rounds. By the guarantee of the multiplicative update method (Theorem 2.5 in [2]), we have that for any $e$ and any $v$

$$
\sum_{t=1}^{T} M(\mathcal{D}^t, p^t) \geq \frac{\ln(1+\epsilon)}{\epsilon} M(\hat{e}, \bar{p}) - \frac{\ln m}{\epsilon}
$$

$$
\sum_{t=1}^{T} M(\mathcal{D}^t, p^t) \geq \frac{\ln(1+\epsilon)}{\epsilon} M(\hat{v}, \bar{p}) - \frac{\ln m}{\epsilon}
$$

Since at time $T$, $w_{\hat{e}}^T = w_{\hat{e}}^0(1+\epsilon)^{M(\hat{e},\bar{p})}$, and $q_{\hat{v}}^T = w_v^0(1+\epsilon)^{M(\hat{v},\bar{p})}$, and the stopping rule ensures that at there exists $e$ or $v$ such that $w_{\hat{e}}^T \geq 1$ or $q_{\hat{v}}^T \geq 1$, we have that either there exists an $e$ such that $M(\hat{e}, \bar{p}) \geq \frac{\ln 1/\delta}{\ln(1+\epsilon)}$ or there exists $v$ such that $M(\hat{v}, \bar{p}) \geq \frac{\ln 1/\delta}{\ln(1+\epsilon)}$. Therefore, by the guarantee of the MWU method, we have that

$$
\sum_{t=1}^{T} M(\mathcal{D}^t, p^t) \geq \frac{\ln 1/\delta}{\epsilon} - \frac{\ln m}{\epsilon}
$$

We now attempt to bound the left-hand-side of the preceding inequality. Note that

$$
\begin{aligned}
M(\mathcal{D}^t, p^t) &= \frac{\sum_e w_{\hat{e}}^t M(\hat{e}^t, p^t) + \sum_v q_{\hat{v}}^t M(\hat{v}^t, p^t)}{\sum_e w_{\hat{e}}^t + \sum_v q_{\hat{v}}^t} \\
&= \frac{A(p^t) \cdot \left( \sum_{e \in \pi^t} w_{\hat{e}}^t / B(e) + \min_{v \in \pi^t} q_{\hat{v}}^t / C(v) \right)}{\sum_e w_{\hat{e}}^t + \sum_v q_{\hat{v}}^t}
\end{aligned}
$$

By the definition of $\pi^t$ and Observation 2, we have

$$M(\mathcal{D}^t, p^t) = \frac{A(p^t)(\sum_{e \in \pi^t} w_{\hat{e}}^t / B(e) + \min_{v \in \pi^t} q_{\hat{v}}^t / C(v))}{\sum_e w_{\hat{e}}^t + \sum_v q_{\hat{v}}^t}$$

$$\leq A(p^t)/A(p^{opt})$$

Combining these inequalities, we get that

$$A(\bar{p})/A(p^{opt}) \geq \sum_{t=1}^{T} M(\mathcal{D}^t, p^t) \geq \frac{\ln 1/(|E| \cdot \delta)}{\epsilon}$$

Fixing any edge $e$, its expert's initial weight is $1/\delta$ and its expert's final weight is at most $1+\epsilon$. Thus, $\bar{p}$ passes at most $B(e) \log_{1+\epsilon}((1+\epsilon)/\delta)$ flow through it. Similarly, for each $v$, at most $C(v) \log_{1+\epsilon}((1+\epsilon)/\delta)$ units of processing are assigned to it. In other words, scaling down all $p^t$ flows by $\log_{1+\epsilon}(1+\epsilon)/\delta$ will result in a feasible flow. Letting $p' = \bar{p}/\log_{1+\epsilon} \frac{1+\epsilon}{\delta}$, we get

$$A(p')/A(p^{opt}) \geq A(\bar{p})/\left(A(p^{opt}) \log_{1+\epsilon} \frac{1+\epsilon}{\delta}\right)$$

$$\geq \frac{\ln(1/(|E| \cdot \delta))}{\epsilon} / \log_{1+\epsilon} \frac{1+\epsilon}{\delta}$$

Taking $\delta = (1+\epsilon)((1+\epsilon)m)^{-1/\epsilon}$, we have that

$$\frac{A(p')}{A(p^{opt})} \geq (1 - \epsilon),$$

giving the promised approximation factor.

Note that in each iteration, we either increase the weight of one $w_{\hat{e}}$ by a factor of $(1 + \epsilon)$, or increase all of the $q_{\hat{v}}$'s on a path $\pi^t$ by a factor of $(1 + \epsilon)$. Since each $w_{\hat{e}}$ and each $q_{\hat{v}}$ can only be increased by such a factor at most $\frac{\ln 1/(|E| \cdot \delta)}{\epsilon}$ times before its weight exceeds 1, the total running time $T$ is bounded by $(|V| + |E|) \frac{\ln 1/(|E| \cdot \delta)}{\epsilon} \cdot T_{sp} = O(|E| \log |V|/\epsilon^2 \cdot T_{sp})$, where $T_{sp}$ is the time it takes Algorithm 2 to compute the optimal path for each the $|D|$ flows. As a single flow takes time $O(|E| + |V| \log V)$ using Fibonacci heaps, we can compute the 2-walk for each of the flows in time $O(|D| \cdot (|E| + |V| \log |V|))$. Thus, the total running time is $O(|D| \cdot |E| \cdot (|E| + |V| \log |V|) \cdot \log^2 |V|/\epsilon^2)$.

## 6    Middlebox Node Purchase Optimization

We now discuss the network design problems mentioned in the introduction. Although such problems can be modeled in multiple ways, we limit our discussion to the case where each vertex $v$ has a potential processing capacity $C$, which can only be utilized if $v$ is "purchased". Flow processed elsewhere can be routed through $v$ regardless of whether or not $v$ is purchased. We give results for both directed and undirected variants of two versions of the network design problem:

1. The *minimization* version (MIN MIDDLEBOX NODE PURCHASE), where the goal is to pick the smallest set of vertices such that all flow is routable.
2. The *maximization* version (BUDGETED MIDDLEBOX NODE PURCHASE), in which we try to maximize the amount of routable flow while subject to a budget constraint of $k$.

Formally, the input to MIN MIDDLEBOX NODE PURCHASE is a (di)graph $G = (V, E)$ with nonnegative costs $q_v$ on its vertices, a potential processing capacity $C : V \to [0, \infty)$, and a collection of $(s_i, t_i)$ pairs with demands $R_i$. The goal is to select a set $T \subseteq V$ of vertices such that all demands are satisfied. BUDGETED MIDDLEBOX NODE PURCHASE is given the same collection of inputs along with a budget integer $k$, and the goal is to route as much of the demand as possible.

All four problems (maximization or minimization, directed or undirected), are **NP**-hard.

## 6.1   Approximation Hardness for Directed Min Middlebox Node Purchase

We now prove that directed MIN MIDDLEBOX NODE PURCHASE is NP-hard to approximate to a factor better than $(1 - \epsilon) \ln n$ by showing an approximation-preserving reduction from SET COVER, a problem already known to have the aforementioned $(1 - \epsilon) \ln n$ hardness [8].

Given a SET COVER instance with set system $\mathcal{S} = \{S_1, S_2, \cdots\}$ and universe of elements $\mathcal{U}$, we create one vertex $v_S$ for each $S \in \mathcal{S}$ and one vertex $w_u$ for each $u \in \mathcal{U}$. Further, we create one source vertex $s$ and one sink vertex $t$, where $t$ demands $|\mathcal{U}|$ units of processed flow from $s$. We add one capacity-$n$ arc from $s$ to each $v_S$, and one capacity-1 arc from each $w_u$ to $t$. We then add a capacity-1 arc from each $v_S$ to $w_u$ whenever $S \ni u$. Finally, we give each $v_S$ vertex $n$ units of processing capacity at a cost of 1 each.

In order for $t$ to get $|\mathcal{U}|$ units of flow, each $w_u$ must get at least one unit of processed flow itself. Thus, at least one of its incoming $v_S$ neighbors must be able to process flow. Therefore, this instance of directed MIN MIDDLEBOX NODE PURCHASE can be seen as the problem of purchasing as few of the $v_S$ vertices so that each $u_W$ vertex has one (or more) incoming $v_S$ vertex. This provides a direct one-to-one mapping between solutions to our constructed instance and the initial SET COVER instance, and the values of the solutions are conserved by the mapping. Therefore, we have an approximation-preserving reduction between the two problems, and directed MIN MIDDLEBOX NODE PURCHASE acquires the known $(1 - \epsilon) \ln n$ inapproximability of SET COVER, summarized in the following result:

**Theorem 2.** *For every $\epsilon > 0$, it is* **NP**-*hard to approximate directed* MIN MIDDLEBOX NODE PURCHASE *to within a factor of* $(1 - \epsilon) \ln n$.

Note that this construction provides the same hardness even when all demands are only to be satisfied up to a $(1 - \delta)$ fraction, showing the asymptotic tightness of the approximation factor in Theorem 3.

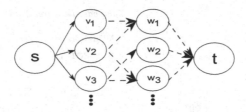

**Fig. 3.** Approximation-preserving reduction from SET COVER and MAX $k$-COVERAGE to directed MIN MIDDLEBOX NODE PURCHASE and directed BUDGETED MIDDLEBOX NODE PURCHASE. Solid edges have infinite capacity, dashed edges have capacity 1. $v_i$ vertices have infinite processing potential, at a cost of 1 each.

## 6.2  Bicriterion Approximation Algorithm for Directed and Undirected Min Middlebox Node Purchase

We first describe an algorithm for directed MIN MIDDLEBOX NODE PURCHASE that satisfies all flow requirements up to a factor of $1 - \delta$ fraction with expected cost bounded by $O(\log n/\delta^2)$ times the optimum.

We begin our approximation algorithm for directed MIN MIDDLEBOX NODE PURCHASE by modifying the 2-walk-based LP formulation with additional variables $x_v$ corresponding to whether or not processing capacity at vertex $v$ has been purchased. We further give a polynomial sized edge-based LP formulation with flow variables $f_i^{1,v}(e)$ and $f_i^{2,v}(e)$ for each commodity $i$, each vertex $v \in V$ and each edge $e \in E$. The variables $f_i^{1,v}(e)$ correspond to the (processed) commodity $i$ flow that has been processed by vertex $v$: these variables describe a flow from $v$ to $t_i$. The variables $f_i^{2,v}(e)$ correspond to the (unprocessed) commodity $i$ flow that will be processed by vertex $v$: these variables describe a flow from $s_i$ to $v$. See Fig. 4 for the full linear program.

Given an optimal solution to this LP, we pick vertices to install processing capacity on by randomized rounding: pick vertex $v$ with probability $x_v$. if $x_v$ is picked, then all flows processed by $v$ are rounded up in the following way: $\hat{F}_i^{j,v}(e) = f_i^{j,v}(e)/x_v$ for all $i \in [|D|], j \in \{1,2\}, e \in E$. If $v$ is not picked, then all flows processed by $v$ are set to zero, i.e. $\hat{F}_i^{j,v}(e) = 0$.

By design, $E[\hat{F}_i^{j,v}(e)] = f_i^{j,v}(e)$. In the solution produced by the rounding algorithm, the total flow through edge $e$ is $\sum_{v \in V} \sum_{i=1}^{|D|} ((\hat{F}_i^{1,v}(e) + \hat{F}_i^{2,v}(e)))$. This is a random variable whose expectation is at most $B(e)$, and is the sum of independent random variables, one for each vertex $v$. The constraints of the LP ensure that if $v$ is selected, then the total processing done by vertex $v$ is at most $C(v)$. Further, the total contribution of vertex $v$ to the flow on edge $e$ does not exceed the capacity $B(e)$, i.e. $\sum_{i=1}^{|D|} (\hat{F}_i^{1,v}(e) + \hat{F}_i^{2,v}(e)) \le B(e)$. Also,

*2-Walk-based formulation:*

MINIMIZE $\sum_{v\in V} q_v x_v$

SUBJECT TO

$$x_v \leq 1 \qquad\qquad \forall v\in V$$

$$p_{i,\pi}=\sum_{v\in\pi} p_{i,\pi}^v \qquad\qquad \forall i\in[|D|],\pi\in P$$

$$\sum_{\pi\in P} p_{i,\pi}\geq R_i \qquad\qquad \forall i\in[|D|]$$

$$\sum_{i=1}^{|D|}\sum_{\substack{\pi\in P\\ \pi\ni e}} p_{i,\pi}\leq B(e) \qquad\qquad \forall e\in E$$

$$\sum_{i=1}^{|D|}\sum_{\pi\in P} p_{i,\pi}^v\leq C(v)x_v \qquad\qquad \forall v\in V$$

$$\sum_{i=1}^{|D|}\sum_{\substack{\pi\in P\\ \pi\ni e}} p_{i,\pi}^v\leq B(e)x_v \qquad\qquad \forall e\in E,v\in V$$

$$\sum_{\pi\in P} p_{i,\pi}^v\leq R_i x_v \qquad\qquad \forall i\in[|D|],v\in V,$$

$$p_{i,\pi}^v\geq 0 \qquad\qquad \forall i\in[|D|],\pi\in P,v\in\pi$$

$$x_v\geq 0 \qquad\qquad \forall v\in V$$

*Edge-based formulation:*

MINIMIZE $\sum_{v\in V} q_v x_v$

SUBJECT TO

$$x_v\leq 1 \qquad\qquad \forall v\in V$$

$$\sum_{e\in\delta^-(u)} f_i^{j,v}(e)=\sum_{e\in\delta^+(u)} f_i^{j,v}(e)$$
$$\forall i\in[|D|],j\in\{1,2\},v\in V,\forall u\in V\setminus\{s_i,t_i,v\}$$

$$\sum_{e\in\delta^-(v)} f_i^{2,v}(e)=\sum_{e\in\delta^+(v)} f_i^{1,v}(e) \quad \forall i\in[|D|],v\in V$$

$$\sum_{v\in V}\sum_{e\in\delta^+(s_i)} f_i^{2,v}(e)\geq R_i \qquad\qquad \forall i\in[|D|]$$

$$\sum_{i=1}^{|D|}\sum_{v\in V}(f_i^{1,v}(e)+f_i^{2,v}(e))\leq B(e) \qquad\qquad \forall e\in E$$

$$\sum_{i=1}^{|D|}\sum_{e\in\delta^-(v)} f_i^{2,v}(e)\leq C(v)x_v \qquad\qquad \forall v\in V$$

$$\sum_{i=1}^{|D|}(f_i^{1,v}(e)+f_i^{2,v}(e))\leq B(e)x_v \qquad\qquad \forall e\in E,v\in V$$

$$\sum_{e\in\delta^+(s_i)} f_i^{2,v}(e)\leq R_i x_v \qquad\qquad \forall i\in[|D|],v\in V$$

$$f_i^{2,v}(e)=0 \qquad\qquad \forall i\in[|D|],v\in V,e\in\delta^-(s_i)$$

$$f_i^{1,v}(e)=0 \qquad\qquad \forall i\in[|D|],v\in V,e\in\delta^+(t_i)$$

$$p_i^{1,v}(e),p_i^{2,v}(e),x_v\geq 0 \qquad\qquad \forall i\in[|D|],v\in V,e\in E$$

**Fig. 4.** Linear programs from Sect. 3 adapted to middlebox placement problems.

the total contribution of vertex $v$ to the commodity $i$ flow is at most $R_i$, i.e.

$$\sum_{e\in\delta^+(s_i)} \hat{F}_i^{2,v}(e)\leq R_i.$$

We repeat this randomized rounding process $t = O(\log(n)/\epsilon^2)$ times. Let $g^k(e)$ denote the total flow along edge $e$, and $h_i^k$ denote the total amount of commodity $i$ flow in the solution produced by the $k$th round of the randomized rounding process. The following lemma follows easily by Chernoff-Hoeffding bounds:

**Lemma 3**

$$Pr\left[\sum_{k=1}^t g^k(e)\geq (1+\epsilon)t\cdot B(e)\right]\leq e^{-t\epsilon^2/3} \qquad\qquad \forall e\in E \qquad (4)$$

$$Pr\left[\sum_{k=1}^t h_i^k\leq (1-\epsilon)t\cdot R_i\right]\leq e^{-t\epsilon^2/2} \qquad\qquad \forall i\in[|D|] \qquad (5)$$

We set $t = O(\log(n)/\epsilon^2)$ so that the above probabilities are at most $1/n^3$ for each edge $e\in E$ and each commodity $i$. With high probability, none of the associated events occurs. The final solution is constructed as follows: A vertex is purchased if it is selected in any of the $t$ rounds of randomized rounding. Thus the expected cost of the solution is at most $t = O(\log(n)/\epsilon^2)$ times the LP optimum. We consider the superposition of all flows produced by the $t$ solutions and scale down the sum by $t(1+\epsilon)$. This ensures that the capacity constraints are satisfied. Note that the vertex processing constraints are also satisfied by the scaled solution. The total amount of commodity $i$ flow is at least $\frac{1-\epsilon}{1+\epsilon} R_i \geq$ $(1 - 2\epsilon)R_i$. Hence we get the following result:

**Theorem 3.** *For directed* MIN MIDDLEBOX NODE PURCHASE, *there is a polynomial time randomized algorithm that satisfies all flow requirements up to factor $1 - \delta$ and produces a solution that respects all capacities, with expected cost bounded by $O(\log(n)/\delta^2)$ times the optimal cost.*

We can modify the LP to simulate the inclusion of an undirected edge with capacity $B(e)$ by adding the constraints for two arcs between its endpoints with capacity $B(e)$ each, as well as an additional constraint requiring that the sum of flows over these two arcs is bounded by $B(e)$. The analysis done above carries through line-by-line, giving the following result.

**Theorem 4.** *For undirected* MIN MIDDLEBOX NODE PURCHASE, *there is a polynomial time randomized algorithm that satisfies all flow requirements up to factor $1 - \delta$ and produces a solution that respects all capacities, with expected cost bounded by $O(\log(n)/\delta^2)$ times the optimal cost.*

### 6.3  Approximation Hardness for Undirected Min Middlebox Node Purchase

We now show an approximation preserving reduction from MIN VERTEX COVER to undirected MIN MIDDLEBOX NODE PURCHASE, proving that the latter problem is **UGC**-hard to approximate within a factor of $2 - \epsilon$ for any $\epsilon > 0$ [18], and **NP**-hard to approximate within a factor of 1.36 [7].

The construction is simple. Given a VERTEX COVER instance with graph $G = (V, E)$, we create an identical graph with each vertex $v$ demanding one unit of processed flow from each of its neighbors, and each edge's capacity is 2. Further, each vertex has $n$ units of processing potential, at a cost of 1. Because the total demand equals the sum of all edge capacities, each unit of flow sent must use exactly one unit of edge capacity, i.e. all flow paths have length exactly one. Thus, the set of solutions exactly corresponds to vertex covers, with one unit of flow going each way across each edge, from source to sink and either to or from its point of processing. The unit costs ensure that the objective value equals the number of vertices picked, and thus that the optimal solution to this undirected MIN MIDDLEBOX NODE PURCHASE instance equals that of the original MIN VERTEX COVER. The conclusion, summarized below, follows.

**Theorem 5.** *Approximating undirected* MIN MIDDLEBOX NODE PURCHASE *is at least as hard as approximating* MIN VERTEX COVER. *In particular, it is* **NP**-*hard to approximate within a factor of* 1.36 *and* **UGC**-*hard to approximate within a factor of* $2 - \epsilon$, *for any $\epsilon > 0$.*

## 6.4   Approximation Algorithm for Directed Budgeted Middlebox Node Purchase

The algorithm here proceeds similarly to that in Sect. 6.2. The LPs we use are the natural maximization variant of those used for the minimization problem, with the added restriction that we only use a 1/2 fraction of the budget. It is easy to see that this additional restriction does not reduce the objective value of the optimal LP solution by more than an 1/2-fraction. We also assume (without loss of generality) that no vertex has cost greater than the budget. The LPs are formulated as follows:

*2-Walk-based formulation:*

MAXIMIZE $\sum_{i=1}^{|D|} \sum_{\pi \in P} p_{i,\pi}$

SUBJECT TO

$\sum_{v \in V} c_v x_v \leq k/2$

$x_v \leq 1$      $\forall v \in V$

$p_{i,\pi} = \sum_{v \in \pi} p_{i,\pi}^v$      $\forall i \in [|D|], \pi \in P$

$\sum_{\pi \in P} p_{i,\pi} \geq R_i$      $\forall i \in [|D|]$

$\sum_{i=1}^{|D|} \sum_{\substack{\pi \in P \\ \pi \ni e}} p_{i,\pi} \leq B(e)$      $\forall e \in E$

$\sum_{i=1}^{|D|} \sum_{\pi \in P} p_{i,\pi}^v \leq C(v) x_v$      $\forall v \in V$

$\sum_{i=1}^{|D|} \sum_{\substack{\pi \in P \\ \pi \ni e}} p_{i,\pi}^v \leq B(e) x_v$      $\forall e \in E, v \in V$

$\sum_{\pi \in P} p_{i,\pi}^v \leq R_i x_v$      $\forall i \in [|D|], v \in V,$

$p_{i,\pi}^v \geq 0$      $\forall i \in [|D|], \pi \in P, v \in \pi$

$0 \leq x_v \leq 1$      $\forall v \in V$

*Edge-based formulation:*

MAXIMIZE $\sum_{v \in V} \sum_{i=1}^{|D|} \sum_{e \in \delta^-(v)} f_i^{2,v}(e)$

SUBJECT TO

$\sum_{v \in V} c_v x_v \leq k/2$

$\sum_{e \in \delta^-(u)} f_i^{j,v}(e) = \sum_{e \in \delta^+(u)} f_i^{j,v}(e)$
     $\forall i \in [|D|], j \in \{1,2\}, v \in V, \forall u \in V \setminus \{s_i, t_i, v\}$

$\sum_{e \in \delta^-(v)} f_i^{2,v}(e) = \sum_{e \in \delta^+(v)} f_i^{1,v}(e)$      $\forall i \in [|D|], v \in V,$

$\sum_{v \in V} \sum_{e \in \delta^+(s_i)} f_i^{2,v}(e) \geq R_i$      $\forall i \in [|D|]$

$\sum_{i=1}^{|D|} \sum_{v \in V} (f_i^{1,v}(e) + f_i^{2,v}(e)) \leq B(e)$      $\forall e \in E$

$\sum_{i=1}^{|D|} \sum_{e \in \delta^-(v)} f_i^{2,v}(e) \leq C(v) x_v$      $\forall v \in V$

$\sum_{i=1}^{|D|} (f_i^{1,v}(e) + f_i^{2,v}(e)) \leq B(e) x_v$      $\forall e \in E, v \in V$

$\sum_{e \in \delta^+(s_i)} f_i^{2,v}(e) \leq R_i x_v$      $\forall i \in [|D|], v \in V$

$f_i^{2,v}(e) = 0$      $\forall i \in [|D|], v \in V, e \in \delta^-(s_i)$

$f_i^{1,v}(e) = 0$      $\forall i \in [|D|], v \in V, e \in \delta^+(t_i)$

$p_i^{1,v}(e), p_i^{2,v}(e), x_v \geq 0$      $\forall i \in [|D|], v \in V, e \in E$

$0 \leq x_v \leq 1$      $\forall v \in V$

If purchasing a single vertex allows us to route a $1/(2 \ln n)$ fraction of the objective value of the above LP, we purchase only this vertex. Otherwise, we can remove the potential for processing at each vertex $v$ with $c_v \geq k/\ln n$ and re-solve the LP to get a solution with objective value at least half as large as before. Thus, from now on we can assume that no $c_v$ exceeds $k/\ln n$ and therefore that the optimal LP solution puts support on at least a $1/\ln n$ fraction of the $x_v$s (at a cost of 2 in our approximation factor). We will call the objective value of this modified linear program OPT$_{\text{LP}'}$.

Again, we pick the vertices on which to install processing capacity on by randomized rounding: each vertex $v$ is picked with probability $x_v$. If $x_v$ is picked, then all flows processed by $v$ are rounded so that $\hat{F}_i^{j,v}(e) = f_i^{j,v}(e)/(4x_v \ln n)$ for all $i \in [|D|], j \in \{1,2\}, e \in E$. If $v$ is not picked, then all flows processed by $v$ are set to zero, i.e. $\hat{F}_i^{j,v}(e) = 0$.

By design, $E[\hat{F}_i^{j,v}(e)] = f_i^{j,v}(e)/(4\ln n)$ and thus the total amount of flow processed, $P$, satisfies $E[P] = E\left[\sum_{v\in V}\sum_{i=1}^{|D|}\sum_{e\in\delta^-(v)}\hat{F}_i^{2,v}(e)\right] = \text{OPT}_{\text{LP}'}/(4\ln n)$. In the solution produced by the rounding algorithm, the total flow through edge $e$ is

$$\sum_{v\in V}\sum_{i=1}^{|D|}((\hat{F}_i^{1,v}(e)+\hat{F}_i^{2,v}(e)). \text{ This sum of random variables is } \hat{B}(e) = B(e)/(4\ln n)$$

in expectation. Letting $g(e)$ denote the flow along edge $e$, standard bounds give

**Lemma 4**

$$Pr\left[g(e) \geq (4\lg n)\cdot\hat{B}(e)\right] \leq e^{-4\ln n} = n^{-4} \qquad \forall e\in E \qquad (8)$$

$$Pr\left[P \leq (1/4)\cdot(1/(4\lg n)\cdot\text{OPT}_{\text{LP}'})\right] \leq e^{-4\ln n} = n^{-4} \qquad \forall e\in E \qquad (9)$$

so by the union bound, with probability higher than $1-1/n$ every edge is assigned $\leq B(e)$ total flow and the amount of flow processed and routed is within a $1/16\ln n$ factor of $\text{OPT}_{\text{LP}'}$.

Finally, by Markov's inequality, the original budget constraint is satisfied with probability at least $1/2$. Combining this with Lemma 4, the algorithm fails with probability at most $1/2+1/n$. Repeating the algorithm $O(\log n)$ times and taking the best feasible solution therefore provides an $\Omega(1/\log n)$ approximation with probability at least $1-1/\text{poly}(n)$. This can be summarized in the following result:

**Theorem 6.** *For directed* BUDGETED MIDDLEBOX NODE PURCHASE, *there is a polynomial-time randomized algorithm producing an $\Omega(1/\log(n))$ approximation.*

We can also apply this algorithm to undirected instances by adding additional constraints the as we did in Sect. 6.2, with the analysis carrying through as before. Thus, we attain the following:

**Theorem 7.** *For undirected* BUDGETED MIDDLEBOX NODE PURCHASE, *there is a polynomial-time randomized algorithm producing an $\Omega(1/\log(n))$ approximation.*

## 6.5 Approximation Algorithm for Undirected Budgeted Middlebox Node Purchase

We now show that the undirected BUDGETED MIDDLEBOX NODE PURCHASE admits a constant-factor approximation algorithm when restricted to a single source $s$. Let $\text{OPT}(G, k)$ denote the value of the optimal solution to an instance with graph $G$ and budget $k$. Our algorithm works by splitting the problem into both a *processing step* and a *routing step*. The algorithm begins by reserving a

1/2 fraction of each edge for use in the processing step and the remaining 1/2 fraction for use in the routing step. Calling the reserved-capacity graphs $G_{\text{proc}}$ and $G_{\text{route}}$, respectively, the algorithm proceeds as follows:

*Processing Step.* A well known fact in capacitated network design is that the maximum amount of flow routable (sans processing) from a set $S \subseteq V$ of source vertices to a single sink forms a monotone, submodular function in $S$ [4]. Although this problem is usually defined in the context of sources that can produce an arbitrary amount of flow (should the network support it), we can bottleneck each source $s_i$ into producing at most some $c_i$ units of flow by replacing it with a pair of vertices connected by a capacity $c_i$ edge, without changing the submodularity of the routable flow function, $f_G(S)$. For the purpose of this lemma, redefining $s$ as our "sink" and the set $P$ of processing nodes as our source set $S$, we immediately attain that the function $f_G(P)$ is submodular, where $P \subset V$ is the set of nodes purchased for processing.

Let $H$ be a copy of $G_{\text{proc}}$ with all edge capacities halved. Because $f_H$ is a submodular function, the problem of using our budget to purchase a set $P \subseteq V$ of processing nodes so to maximize $f_H(P)$ is simply an instance of a monotone, submodular maximization subject to knapsack constraints. Such problems are known to admit simple $(1 - 1/e)$-approximation algorithms [30]. Let $P(H, k)$ be the optimal solution to this *processable flow problem* on $H$ with budget $k$ and $\text{ALG}_1(H, K)$ denote the value of the solution found by our algorithm. Because $P(H, k)$ is an upper bound on $\text{OPT}(H, k)$ (indeed, the former is simply an instance of the former without the need to account for post-processing routing), the $(1 - 1/e)$ approximation we get has value at least equal to $(1 - 1/e)$ times the value of $\text{OPT}(H, k)$. In particular

$$
\begin{aligned}
\text{ALG}_1(H, k) &\geq (1 - 1/e)P(H, k) \\
&\geq (1 - 1/e)\text{OPT}(H, k) \\
&\geq (1 - 1/e)(1/2)\text{OPT}(G_{\text{proc}}, k) \\
&\geq (1 - 1/e)(1/2)(1/2)\text{OPT}(G, k) \\
&= (1 - 1/e)/4 \cdot \text{OPT}(G, k)
\end{aligned}
$$

Further, because our solution only uses at most half of the capacity of any edge in $G_{\text{proc}}$, we can use the remaining, unused half of the capacities to route all flow we managed to process back to $s$.

*Routing Step.* All flow residing in $s$ after the end of the processing step is already processed, all of it can be routed directly to the sinks using the 1/2 fraction of edge capacities we reserved for $G_{\text{route}}$. Because multiplying all edge capacities by 1/2 reduces the amount of routable flow by the same (multiplicative) amount, we can route at least $(1/2)\min(\text{ALG}_1(H, k), \text{MAXFLOW}_G(s, t))$ units of the processed flow from $s$ to $t$. As $\text{MAXFLOW}_G(s, t)$ is a (trivial) upper bound on $\text{OPT}(G, k)$, this means we can route at least $(1/2)(1-1/e)/4\text{OPT}(G, k)$ units of the processed flow from $s$ to the sinks, giving a $(1-1/e)/8 > .078$ approximation algorithm.

Thus, we get the following theorem:

**Theorem 8.** *For undirected* BUDGETED MIDDLEBOX NODE PURCHASE *with a single source, there is a deterministic polynomial time algorithm that produces a solution that can route at least* $(1 - 1/e)/8 \approx .078$ *times the optimal objective solution.*

## 6.6 Approximation Hardness for Directed Budgeted Middlebox Node Purchase

We now prove that directed BUDGETED MIDDLEBOX NODE PURCHASE is NP-hard to approximate to a factor of $1 - 1/e + \epsilon$. To show this, we reduce from MAX K-COVER, which is known to have the same hardness result [12].

Given a MAX K-COVER instance with set system $\mathcal{S}$ and universe of elements $\mathcal{U}$, we create one vertex $v_S$ for each $S \in \mathcal{S}$ and one vertex $w_u$ for each $u \in \mathcal{U}$. Further, we create one source vertex $s$ and one sink vertex $t$, where $t$ demands $|\mathcal{U}|$ units of processed flow from $s$. We add one capacity-$n$ arc from $s$ to each $v_S$, and one capacity-1 arc from each $w_u$ to $t$. We then add a capacity-1 arc from each $v_S$ to $w_u$ whenever $S \ni u$. Finally, we give each $v_S$ vertex $n$ units of processing capacity at a cost of 1 each. The budget for the instance is $k$ – the same as the budget for the MAX-K-COVER instance. A diagram of the reduction is given in Fig. 3.

When flow is routed maximally, each $w_u$ contributes 1 unit of flow to the total $s - t$ flow if and only if it has a neighbor $v_S$ that was chosen to be active. Otherwise, this vertex does not help contribute towards the $s - t$ flow. Thus, this instance of directed BUDGETED MIDDLEBOX NODE PURCHASE can be seen as the problem of buying $k$ different $v_S$ vertices so to maximize the number of distinct $w_u$ vertices to which they are adjacent. Thus, there is a direct one-to-one mapping between solutions to our constructed instance and the initial MAX K-COVER instance, and the values of the solutions are conserved by the mapping. Therefore, we have an approximation-preserving reduction between the two problems, and directed BUDGETED MIDDLEBOX NODE PURCHASE acquires the known $(1 - 1/e + \epsilon)$ inapproximability of MAX K-COVER. The result can be summarized as follows

**Theorem 9.** *For every* $\epsilon > 0$, *it is* **NP**-hard *to approximate directed* BUDGETED MIDDLEBOX NODE PURCHASE *to within a factor of* $1 - 1/e + \epsilon$.

## 6.7 Approximation Hardness for Undirected Budgeted Middlebox Node Purchase

We show that for some fixed $\epsilon_0 > 0$, the undirected version of BUDGETED MIDDLEBOX NODE PURCHASE is **NP**-hard to approximate within a factor of $1 - \epsilon$, implying that the problem does not admit a PTAS unless **P** = **NP**. We make no attempt to maximize the value $\epsilon_0$.

We show this hardness by reducing from MAX BISECTION on degree-3 graphs, shown to be hard to approximate within a factor of .997 in [3][1]. Let $G = (V, E)$ be the input to the degree-3 MAX BISECTION instance. For each $v_i \in V$, create two vertices, $u_i$ and $w_i$, joined by an edge with capacity 3. We also add a capacity-1 edge between $u_i$ and $u_j$ whenever $v_i$ and $v_j$ are adjacent in $G$. Each $w_i$ vertex demands 3 units of flow from every $u_j$ (including when $i = j$). Further, every $u_i$ vertex can be given $3|V|$ units of processing capacity (or, equivalently, $\infty$ units) at a cost of 1, and the instance's budget is set to $|V|/2$.

The intuition behind the construction is as follows. With a budget of $|V|/2$, we can purchase exactly half of the $u_i$ vertices (and all budget is used up without loss of generality); our bisection will be between the purchased $u_i$s and the unpurchased ones. Let $b$ be the number of edges in any such bisection. Each $w_i$ adjacent to a purchased $u_i$ can have 3 units of its demand satisfied by flow originating from and processed by $u_i$, and the only edge connecting $w_i$ to the rest of the graph ensures $w_i$ can never receive more than 3 units of flow regardless. Thus, such $w_i$s are maximally satisfied, and contribute $3|V|/2$ units to our objective value. The remaining $w_i$s must have their processed flow routed to them via edge via the $b$ capacity-1 edges in the bisection (and, indeed, every edge in the bisection will carry 1 unit of flow when routed optimally, as witnessed by the solution where each unprocessed $u_i$ receives flow on each cut-edge and routes it directly to $w_i$), so the total amount of demand satisfied by the $w_i$ adjacent to unprocessed vertices is exactly $b$, so the objective value of a solution with $b$ edges in the bisection is exactly $3|V|/2 + b$.

Let $b_{\text{OPT}}$ denote the number of edges cut by the optimal bisection. It is a well-known fact that $b_{\text{OPT}} \geq |E|/2 = 3|V|/4$. By the theorem of [3] it is **NP**-hard to distinguish instances with $3|V|/2 + b_{\text{OPT}}$ units of satisfiable demand from those with only $3|V|/2 + (1 - .003)b_{\text{OPT}}$, giving an inapproximability ratio of

$$\frac{3|V|/2 + (1 - .003)b_{\text{OPT}}}{3|V|/2 + b_{\text{OPT}}} = 1 - \frac{.003 b_{\text{OPT}}}{3|V|/2 + b_{\text{OPT}}}$$

$$= 1 - \frac{.003}{3|V|/(2b_{\text{OPT}}) + 1}$$

$$\leq 1 - \frac{.003}{3|V|/(2 \cdot 3|V|/4) + 1}$$

$$= 1 - \frac{.003}{2 + 1}$$

$$= .999$$

This calculation is summarized in the following result:

**Theorem 10.** *It is* **NP**-*hard to approximate undirected* BUDGETED MIDDLE-BOX NODE PURCHASE *to within a factor better than* .999.

---

[1] To be precise, this paper shows the aforementioned hardness for MAX CUT. A simple approximation preserving reduction from MAX CUT to MAX BISECTION can be derived by looking at maximum cuts of the graph formed by 2 disjoint copies of the MAX CUT instance graph.

# 7   Related Work

*Network Function Optimization.* In software-defined networking, SIMPLE [24] and FlowTags [11] take advantage of switches with fine-grained rule support. Both approaches focus on how to use the constrained TCAM size, a hardware limitation to support fine-grained policy. Neither approach attempts to solve the joint optimization of the capacity constraints for both servers and switches. Slick [1] offers a high-level control program that specifies custom processing on precise subset of flows. It also assumes the server processing power is heterogeneous, and uses heuristic approaches for the underlying placement, routing, and steering.

*Network Function Consolidation.* CoMB [28] and Click [21] both consolidate network functions into applications or a VM images, and consider server machines that can each run multiple instances of different network functions. Both focus on improving the performance on single nodes, and treat network functions homogeneously. Neither covers a network-wide optimization.

*Network Function Migration and Reroute.* OpenNF [14] and Split-Merge [25] leverage the SDN controller to manage the network function's state migration and the network function's flow migration. Both focus on reallocating resources and rerouting flows when either a node or a link is over-utilized. While their solution focuses on fixing congestion when it occurs, ours focuses on figuring out how to avoid congestion in the first place.

*Network Function Online Request Model.* Recently, Even, Medina, and Patt-Shamir [9] studied an online request admission problem in the same multi-commodity flow with processing setting that we study. In their work, requests arrive online and specify a processing pipeline for flow between a source and sink; intermediate nodes in the pipeline may be any subset of nodes in the underlying graph. The goal is to accept as many such flow requests as possible while ensuring that accepted requests are assigned flow paths that satisfy capacity constraints. In this setting, the authors show an $O(k \log(kn))$-competitive online algorithm for instances with length-$k$ pipelines.

*Routing and Middlebox optimization.* A couple of recent papers consider approximation algorithms for path computation and service placement [10] and Service Chain and Virtual Network Embeddings [26,27]. Both papers use randomized rounding of a linear programming relaxation of the problem. Both of these works differ from our paper in that packets between demand pairs are not splittable, and thus must be sent along *paths* rather than *flows*. Other recent papers provide approximation algorithms for variants of MIN MIDDLEBOX NODE PURCHASE with no hard edge constraints [6,20]. In [20], the authors independently derive the same SET COVER-based hardness construction for their problem variant.

# References

1. Anwer, B., Benson, T., Feamster, N., Levin, D.: Programming Slick network functions. In: Proceedings of Symposium on SDN Research, June 2015
2. Arora, S., Hazan, E., Kale, S.: The multiplicative weights update method: a meta-algorithm and applications. Theor. Comput. **8**(1), 121–164 (2012)
3. Berman, P., Karpinski, M.: On some tighter inapproximability results (extended abstract). In: Wiedermann, J., van Emde Boas, P., Nielsen, M. (eds.) ICALP 1999. LNCS, vol. 1644, pp. 200–209. Springer, Heidelberg (1999). https://doi.org/10. 1007/3-540-48523-6_17
4. Chakrabarty, D., Krishnaswamy, R., Li, S., Narayanan, S.: Capacitated network design on undirected graphs. In: Raghavendra, P., Raskhodnikova, S., Jansen, K., Rolim, J.D.P. (eds.) APPROX/RANDOM -2013. LNCS, vol. 8096, pp. 71–80. Springer, Heidelberg (2013). https://doi.org/10.1007/978-3-642-40328-6_6
5. Chiosi, M., et al.: Network functions virtualisation: introductory white paper. In: SDN and OpenFlow World Congress, October 2012
6. Cohen, R., Lewin-Eytan, L., Naor, J.S., Raz, D.: Near optimal placement of virtual network functions. In: IEEE Conference on Computer Communications (INFO-COM), pp. 1346–1354. IEEE (2015)
7. Dinur, I., Safra, S.: On the hardness of approximating minimum vertex cover. Ann. Math. **162**, 439–485 (2005)
8. Dinur, I., Steurer, D.: Analytical approach to parallel repetition. In: Proceedings of the Annual ACM Symposium on Theory of Computing, pp. 624–633. ACM, New York (2014). https://doi.org/10.1145/2591796.2591884. http://doi.acm.org/ 10.1145/2591796.2591884
9. Even, G., Medina, M., Patt-Shamir, B.: Competitive path computation and function placement in SDNs. arXiv preprint arXiv:1602.06169 (2016)
10. Even, G., Rost, M., Schmid, S.: An approximation algorithm for path computation and function placement in SDNs. In: Suomela, J. (ed.) SIROCCO 2016. LNCS, vol. 9988, pp. 374–390. Springer, Cham (2016). https://doi.org/10.1007/978-3-319-48314-6_24
11. Fayazbakhsh, S.K., Chiang, L., Sekar, V., Yu, M., Mogul, J.C.: Enforcing network-wide policies in the presence of dynamic middlebox actions using flowtags. In: 11th USENIX Symposium on Networked Systems Design and Implementation (NSDI 2014), pp. 543–546. USENIX Association, Seattle, April 2014. https://www.usenix. org/conference/nsdi14/technical-sessions/presentation/fayazbakhsh
12. Feige, U.: A threshold of ln n for approximating set cover. J. ACM (JACM) **45**(4), 634–652 (1998)
13. Forrest, J.: Clp: Coin-or linear program solver. In: DIMACS Workshop on COIN-OR, pp. 17–20, July 2006
14. Gember-Jacobson, A., et al.: OpenNF: enabling innovation in network function control. In: Proceedings of the ACM Conference on SIGCOMM, pp. 163–174. ACM (2014). https://doi.org/10.1145/2619239.2626313. http://doi.acm.org/ 10.1145/2619239.2626313
15. Heorhiadi, V., Reiter, M.K., Sekar, V.: Accelerating the development of software-defined network optimization applications using SOL. arXiv preprint arXiv:1504.07704 (2015)
16. Heorhiadi, V., Reiter, M.K., Sekar, V.: Simplifying software-defined network optimization using sol. In: 13th USENIX Symposium on Networked Systems Design and Implementation (NSDI 2016), pp. 223–237. USENIX Association,

Santa Clara, March 2016. https://www.usenix.org/conference/nsdi16/technical-sessions/presentation/heorhiadi

17. Jin, Y., Wen, Y., Westphal, C.: Towards joint resource allocation and routing to optimize video distribution over future internet. In: IFIP Networking Conference (IFIP Networking) 2015, 1–9 May 2015. https://doi.org/10.1109/IFIPNetworking.2015.7145311

18. Khot, S., Regev, O.: Vertex cover might be hard to approximate to within 2-$\varepsilon$. J. Comput. Syst. Sci. **74**(3), 335–349 (2008)

19. Li, X., Qian, C.: A survey of network function placement. In: 13th IEEE Annual Consumer Communications Networking Conference (CCNC), pp. 948–953, January 2016. https://doi.org/10.1109/CCNC.2016.7444915

20. Lukovszki, T., Rost, M., Schmid, S.: Approximate and incremental network function placement. J. Parallel Distrib. Comput. **120**, 159–169 (2018)

21. Martins, J., et al.: Clickos and the art of network function virtualization. In: 11th USENIX Symposium on Networked Systems Design and Implementation (NSDI 2014), pp. 459–473. USENIX Association, April 2014. https://www.usenix.org/conference/nsdi14/technical-sessions/presentation/martins

22. OPNFV: OPNFV: an open platform to accelerate NFV, Linux Foundation. https://www.opnfv.org/

23. Orlowski, S., Wessäly, R., Pióro, M., Tomaszewski, A.: Sndlib 1.0 – survivable network design library. Networks **55**(3), 276–286 (2010)

24. Qazi, Z.A., Tu, C.C., Chiang, L., Miao, R., Sekar, V., Yu, M.: SIMPLE-fying middlebox policy enforcement using SDN. In: Proceedings of ACM SIGCOMM, pp. 27–38. ACM (2013). https://doi.org/10.1145/2486001.2486022. http://doi.acm.org/10.1145/2486001.2486022

25. Rajagopalan, S., Williams, D., Jamjoom, H., Warfield, A.: Split/merge: system support for elastic execution in virtual middleboxes. In: Presented as Part of the 10th USENIX Symposium on Networked Systems Design and Implementation (NSDI 2013), pp. 227–240. USENIX, Lombard (2013). https://www.usenix.org/conference/nsdi13/technical-sessions/presentation/rajagopalan

26. Rost, M., Schmid, S.: Charting the complexity landscape of virtual network embeddings. In: IFIP Networking, May 2018. http://eprints.cs.univie.ac.at/5580/

27. Rost, M., Schmid, S.: Virtual network embedding approximations: leveraging randomized rounding. In: IFIP Networking, May 2018. http://eprints.cs.univie.ac.at/5579/

28. Sekar, V., Egi, N., Ratnasamy, S., Reiter, M.K., Shi, G.: Design and implementation of a consolidated middlebox architecture. In: Proceedings of the 9th USENIX Conference on Networked Systems Design and Implementation, NSDI 2012, p. 24. USENIX Association (2012). http://dl.acm.org/citation.cfm?id=2228298.2228331

29. Sherry, J., Hasan, S., Scott, C., Krishnamurthy, A., Ratnasamy, S., Sekar, V.: Making middleboxes someone else's problem: network processing as a cloud service. In: Proceedings of the ACM SIGCOMM 2012 Conference on Applications, Technologies, Architectures, and Protocols for Computer Communication, SIGCOMM 2012, pp. 13–24. ACM (2012). https://doi.org/10.1145/2342356.2342359. http://doi.acm.org/10.1145/2342356.2342359

30. Sviridenko, M.: A note on maximizing a submodular set function subject to a knapsack constraint. Oper. Res. Lett. **32**(1), 41–43 (2004)

31. Uhlig, S., Quoitin, B., Lepropre, J., Balon, S.: Providing public intradomain traffic matrices to the research community. ACM SIGCOMM Comput. Commun. Rev. **36**(1), 83–86 (2006)

# On-Line Big-Data Processing for Visual Analytics with Argus-Panoptes

Panayiotis I. Vlantis[✉] and Alex Delis[✉]

University of Athens, 15703 Athens, Greece
{panosv,ad}@di.uoa.gr

**Abstract.** Analyses with data mining and knowledge discovery techniques are not always successful as they occasionally yield no actionable results. This is especially true in the Big-Data context where we routinely deal with complex, heterogeneous, diverse and rapidly changing data. In this context, visual analytics play a key role in helping both experts and users to readily comprehend and better manage analyses carried on data stored in *Infrastructure as a Service* (*IaaS*) cloud services. To this end, humans should play a critical role in continually ascertaining the value of the processed information and are invariably deemed to be the instigators of *actionable* tasks. The latter is facilitated with the assistance of sophisticated tools that let humans interface with the data through *vision* and *interaction*. When working with Big-Data problems, both scale and nature of data undoubtedly present a barrier in implementing responsive applications. In this paper, we propose a software architecture that seeks to empower Big-Data analysts with visual analytics tools atop large-scale data stored in and processed by *IaaS*. Our key goal is to not only yield *on-line* analytic processing but also provide the facilities for the users to effectively interact with the underlying *IaaS* machinery. Although we focus on hierarchical and spatiotemporal datasets here, our proposed architecture is general and can be used to a wide number of application domains. The core design principles of our approach are: *(a)* On-line processing on cloud with **Apache Spark**. *(b)* Integration of *interactive programming* following the notebook paradigm through **Apache Zeppelin**. *(c)* Offering robust operation when data and/or schema change on the fly. Through experimentation with a prototype of our suggested architecture, we demonstrate not only the viability of our approach but also we show its value in a use-case involving publicly available crime data from United Kingdom.

**Keywords:** Visual analytics · Interactive programming ·
Big-Data processing · Apache Spark · *IaaS* Infrastructures

## 1 Introduction

Datasets used by Big-Data systems and applications are characterized by their complexity, heterogeneity, instant growth, and frequently, noise. These characteristics do affect the quality of automatic analyses performed in a negative way

© Springer Nature Switzerland AG 2019
Y. Disser and V. S. Verykios (Eds.): ALGOCLOUD 2018, LNCS 11409, pp. 102–117, 2019.
https://doi.org/10.1007/978-3-030-19759-9_7

and occasionally, render analyses results to be of either limited or no value at all [11]. By providing appropriate tools, *visual analytics* can help users manually interact with datasets, proceed in an highly exploratory manner and shift the focus of the analyses as the occasion calls along the way [8,16,19]. However, the traditional use of visualization techniques on large scale datasets does become prohibitive as the volume of the underlying data grows [3]. To address this challenge, we have to adopt contemporary cloud-based computing environments that can accommodate voluminous data by incrementally enlarging the computing cluster (i.e., horizontal scaling).

Apache Spark of the Hadoop ecosystem offers a plausible choice as it can scale up when it comes to non-transactional data [21]. However, the use of Spark as the underlying processing engine of applications calling for high responsiveness is not an obvious choice. Spark introduces inherent latencies that cannot be avoided, only mitigated. Clearly, a number of compensating mechanisms have to be introduced to address this issue. Moreover, to further enhance the user experience, we advocate the integration of *interactive programming* in such Big-Data environments. This choice may greatly assist the work of scientist(s) as it offers versatility in handling data and timely decision making during the early exploratory phase that accompanies working with unfamiliar datasets. It is worth mentioning however that, in our case, by introducing the interactive programming paradigm in this Big-Data context, we cannot exploit pre-computation techniques; there are no guaranties as far as the stability of the data is concerned and the data schema remains highly volatile.

In this paper, we propose a software architecture that helps users effectively interact with underlying *IaaS* stored data, manipulate information using Spark and last but not least, enable *on-line* analytic processing via interactive programming. We mitigate Spark-emanating overheads through the introduction of *(1) visualization-chunks*, variable-size granules containing elements shipped over the network and ultimately rendered and presented to users, *(2) schema convergence* techniques enabling the seamless transition among different data schemas used across multiple iterations in run-time, and *(3)* deployment and intensive use of *caching* at all levels of our software architecture. The aforementioned features can work in tandem and take advantage of hierarchical datasets that we have worked with [9].

Figure 1 depicts the salient features of our proposed architecture. It is decoupled in two key parts: a cloud-based *IaaS* as well as a client-side component. At the server side, Apache Spark is used as the Big-Data processing engine accepting requests from Zeppelin and *Visual Analytics Server (VA-Server)*. The Spark-server(s) undertake the actual computation and/or management of the stored datasets. Zeppelin is the interactive programming "notebook platform" essentially offering a Web-interface accessible to users through a browser. This notebook-style facility allows users to execute task in a way reminiscent to that of shell scripting and is the prime tool for direct interaction with the Spark engine and subsequently, for manipulating its *IaaS*-stored data. The *Visual Analytics*

*Server* undertakes the central role of coordinating operations among all cloud components and maintains bi-directional `WebSocket` channels with the client side.

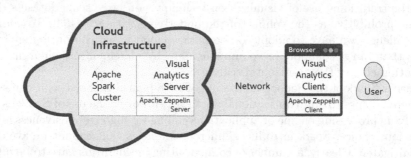

**Fig. 1.** `Argus-Panoptes` architecture for on-line Big-Data visual analytics.

The client-side is a *JavaScript* Web application that runs on the user's browser and carries out the functionality of the *Visual Analytics Client*. The latter renders all server-emanating visualization chunks and accepts user-instigated requests through mouse interaction. All parts of the *JavaScript* application are `React` components. Subsequently, they have to be adapted on a per-case basis however, the use of `React` places strong emphasis on the reusability of components already developed. The `Apache Zeppelin` interface window found in the client-side, allows for the on-line interactive execution of explicit user requests as well as the display of execution outcome which took place in its `Zeppelin`-server counterpart.

In realizing our proposed architecture, our core design principles have been: *(a)* our visual analytics application to carry out its processing on-line on a `Spark`-cluster; hence, our application is independent of the volume of data utilized. *(b)* dataset filtering, joining with others, and transformations are carried out through interactive programming and independently occur from all *VA* aspects. In this regard, users can simultaneously manipulate data through an `Apache Zeppelin` browser-window while at the same time the *VA Client* interface remains fully operational. We argue that this two-pronged approach can effectively overcome the challenges of pursuing visual analytics on Big-Data while at the same time, it yields the basis for overcoming the occasional sluggish response times. Our approach seeks to empower the work of domain experts working along with Big-Data analysts to gain insights and a better understanding through visualization in sophisticated hierarchical datasets.

We have produced a fully-functional prototype, `Argus-Panoptes`[1], that has served as the means to explore a number of Big-Data use-cases We have published

---

[1] Argus-Panoptes is a figure from Greek mythology, it was an "all-seeing" giant having a watchman role.

the source code repository[2] to allow further testing with other datasets and comparisons with similar tools. In this paper, we discuss the effectiveness of our prototype using a real-world Big-Data dataset pertinent to crime incidents curated and published from *UK Home Office* [9]. The dataset in question maintains spatiotemporal records of incidents since late 2015. The rest of the paper is organized as follows: Sect. 2 discusses related work and Sect. 3 presents the rationale for the design of our architecture. Section 4 outlines the architectural components and their interaction. Section 5 briefly discusses our use-case and Sect. 6 provides concluding remarks.

## 2 Related Work

There has been a flurry of research activity recently in the areas of visualization for Big-Data, visual analytics, and visualization recommendation systems [2,4,12,15,18,20]. Apache Zeppelin [1], Cloudera Hue [2] and Jupyter [10] are open-source initiatives that offer *built-in* visualization functionalities, commonly used in Big-Data exploration. All three systems allow their users to "send in" either high-level source code such as *Python* and *Scala* modules or dispatch SQL-queries for execution to Spark. Visualization of pertinent results is realized with the help of built-in visualization libraries [13]. The main difference between our platform and the aforementioned projects is that we strive to offer user a more immersive experience in visual exploration without calling for continual editing of source code or SQL statements in order to bring about changes in rendered visualizations. In contrast, we let users directly interact with the visualizations produced and the respective interface (i.e., *VA Client*). Moreover, we do not strip the ability to directly manipulate data through high-level source code as our platform does also integrate Zeppelin in its components.

In the area of visual analysis of high-dimensional datasets, *Visualization Recommendation (VisRec)* systems offer a novel approach as they suggest feasible visualizations without major user involvement. These systems, automatically designate and interactively suggest visualization choices for specific tasks at hand. In this regard, such recommendations are particularly useful during the initial phase(s) of exploratory analyses through the creation of a series of alternative visualizations. *VisRec* systems operate by performing pre-computations to analyze the dataset during the off-line phase and examine the large space of possible visualization combinations during the on-line phase [18]. While these systems are primed for high-dimensional datasets, their computational intensive on-line phase may make them unsuitable for large scale data without extensive use of *sampling*.

The use of an *RNN* neural-network is advocated in [4] as the means to help novice users start with visualization. The *RNN* network "examines" a corpus of human-created visualization configurations known so far and along with the schema of the used data it automatically generates a json-based visualization

---

[2] Source code repository is available at: https://github.com/panayiotis/visual_analytics.

configuration. The latter is ultimately consumed by JavaScript libraries to display the expected output. In [15], the ZQL-language is proposed as the vehicle to help users designate visual patterns. Such patterns have been extracted from diverse disciplines including biology, engineering, meteorology and commerce. Although this is certainly a novel approach, the number of ZQL-produced possible visualizations remains very high, yielding a somewhat questionable route when it comes to dealing with Big-Data. Evidently, the overall ZQL process presents overheads that would be hard to overcome when on-line processing is sought.

The *imMens* project seeks to provide interactive visualization for Big-Data in real-time [12]. Similarly to our approach, data binning plays an important role as it is the key technique to attain dimensionality reduction. However, *imMens* overall operations are founded on the concept of *pre-computation* of all data-tiles. This pre-computation occurs in an off-line phase and the respective results are made available at runtime to help user fulfill her/his visual analytics tasks. In contrast, our approach performs the respective data tiling by incrementally and dynamically producing json *chunks* that can be created on-the-fly empowering so the on-line mode of operation.

## 3   Argus-Panoptes Design Principles

We intend on furnishing a software architecture that best serves the merged operations of visual analytics and Big-Data analysis. In doing so, Argus-Panoptes should not be restricted by the scale of data while at the same time, the architecture should incorporate core visual analytics principles and should display satisfactory responsiveness. Our design leans towards accommodating the experienced user-base as we would like to create a highly-versatile and efficient architecture. In this context, Argus-Panoptes maintains an open aggregation and exposes internal components/subsystems to the user. Our design addresses the misgivings of contemporary systems that offer visual analytics on Big-Data today. We aspire to address the following 6 design principles while designing Argus-Panoptes:

• **On-Line Big-Data Processing:** weaving a platform such as Spark for data processing along with the visual analytics application atop is not a straight-forward effort. This is due to the fact that it might take several seconds for the Spark-cluster to respond to even a quick look-up query. In contrast, typical responses in a visual interface are expected to be within the 200 ms range. Should we be able to *bridge* the above performance gap, we are to successfully address the design principle in question. Instead of downsizing data through sampling, we advocate *elasticity* of Spark-workers requested by the Argus-Panoptes user. We take advantage of the fact that larger datasets call for horizontal scaling of the cluster as applicable operations (i.e., filtering, aggregations etc.) are highly-parallelizable. By delegating all data operations to the Spark-cluster, Argus-Panoptes reaps the following benefits: (*1*) the *VA* application (both client and server components) becomes yet another component in the Hadoop ecosystem. Consequently, a large number of tools can be readily integrated into our processing pipeline. (*2*) users do not have to code in order to export datasets to specific formats required by the *VA* application. The *VA* application has direct access to `DataFrames` in memory as it functions along with `Spark`. Hence, data pre-processing, cleaning, and *VA* transformations can all be instigated through the `Spark` programming `API` using `Scala, R, Python,` or `Java`.

• **Interactive Programming:** for the user to enjoy the maximum benefit while interacting with our architecture, we introduce interactive notebook systems. Such systems include `Zeppelin` [1], `Jupyter` [10] and the proprietary `Databricks`. In general, they all offer a Web-interface in which a user may write code split into *paragraphs* or blocks each pertaining to a specific job or set of jobs. Paragraphs can be executed either sequentially or individually. Individual execution means that if the notebook crashes in the $k^{th}$ paragraph for example, after it has performed some expensive computations, the user can edit the code on the $k^{th}$ paragraph and continue execution from that point on. This paradigm of code writing and execution is very much desired in our architecture for it facilitates the work of the analyst.

• **Robust Data/Schema Manipulations:** to attain flexibility, the *VA* application has to operate under uncertainty as far as the current data and its schema is concerned. Robustness of this type is particularly desired as `Argus-Panoptes` deploys interactive programming. In a normal operational work-flow, a user simply manipulates a dataset by either adding or removing features (columns). In erroneous circumstances, issues that may ensue include: *(1)* The *VA Client* does not show any data for a user has simply committed a mistake; here, the respective piece of code has to be revised and the query cycle to be repeated. *(2)* The *VA Client* becomes unable to cope with a voluminous visualization chunk consisting in the order of more than $1M$ rows; the user has to reload the browser tab and start over. In all above cases, we should stress that the *VA Client* functionality is desired to remain strictly stateless.

• **Eliminate Unnecessary Re-computations via Caching:** It often takes `Spark` several seconds to compute a visualization chunk. This overhead can be reduced but can not be avoided if identical chucks are requested time and again. Thus, `Spark` re-computations should and are avoided in our architecture through the adoption of 3 levels of caching: *(1)* `Apache Spark`: when a task is dispatched by the *VA Server*, `Spark` can avoid execution should it maintain a pool of executed so far jobs. If a `json` file is identified as existing in this pool, its re-computation is bypassed. *(2)* the *VA Server* tracks with the help of an `SQLite` database all chunks produced so far and in this manner, contact with `Spark` is successfully prevented. This caching layer is possible to return stale data due to this reason, a cache invalidation has to be provisioned. *(3)* the *VA Client* maintains in-memory a limited number of chunks[3] thus requests for fetching locally existing chunks from *VA Server* are eliminated.

---

[3] Around 200 MB in total.

- **Expose System Internals to User:** As `Argus-Panoptes` operation is intended for the experienced user, the system should make available both internal control mechanisms and system information. Such control mechanism is the invalidation of caches at either `Spark` or *VA Server* components. By adopting such a design choice, we avoid error-prone implementation issues that bear little if any significance to the visual analytics domain. Moreover, we argue that making available `Spark`/*VA* server real-time status information is helpful to the experienced user.

- **Promote Visualization Component Reusability with `React`:** Creating a new *VA* application calls for a substantial amount of work most of which is geared towards the development of visualization elements of the interface. This is due to the fact that the effectiveness of a *VA* platform highly depends on tailor-made visualization components. The `React`-framework [7] promotes and provides reuse of `JavaScript`-components in order to built interfaces across multiple *VA* applications. As we strive for `Argus-Panoptes` to not be a one-size-fits-all solution, we resort to the accumulated set of `React` reusable components to rapidly create and/or adapt `VA` customized interfaces.

## 4   The `Argus-Panoptes` System

`Argus-Panoptes` follows the interactive work-flow of operations that filter, aggregate, summarize and help visually explore diverse aspects of datasets under examination. Figure 2 depicts the interface that the *VA Client* of the system realizes. The UI panel consists of two portions: the first is the `Zeppelin` browser window that serves as the means to interact with `Argus-Panoptes` when it comes to launching of work-flow tasks. On the right side of Fig. 2, the *VA Client* browser window displays chart-based outcomes and generated map-related elements. The latter depicts generated graphs that help demonstrate trends and assist users gain insight with regards to investigated datasets.The functionality of `Argus-Panoptes` is built around these three concepts which make the architecture feasible: *Schema Convergence*, *Data Binning* and *Visualization Chunks*. In particular:

- *Schema Convergence* allows the architecture to be fault-tolerant when the schema of the examined dataset is being actively manipulated. This mechanism is always utilized when the *VA Server* ships code to `Spark` for execution. When `Spark` engine receives a *VA Client* request, it compares the schema embedded in the request with the current schema of the dataset. Should discrepancies be identified, `Spark` deals with convergence so that every element on the stack of the system "perceives" a consistent view. In this process, `Spark` imposes no restrictions on the user requests as those are often consistent with a state of the data at an earlier point in time. The data requests instigated from the *VA Client* are predominantly based on what the user has seen last. For instance, if there are new columns in the dataset, they will be included in the converged schema; the same is true when certain columns get dropped.

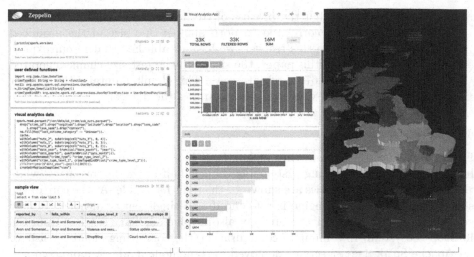

| Apache Zeppelin browser window | Visual Analytics Client browser window |

**Fig. 2.** *VA client* UI.

This schema convergence mechanism disengages the architecture from having to deal with state information in the request-response cycle. Evidently, the user can replace the entire dataset through the UI and the virtual analytics platform will continue to operate trouble-free. Such event is an excellent example of a case where users should explicitly initiate a cache invalidation procedure from the *VA Client* to avoid data inconsistencies.

- **Data Binning** is heavily used in work-flow processing carried out by our platform. It is a critical mechanism to attain graceful dimensionality reduction for discretizing continuous features; oftentimes, data gets summed before sent for visualization to the *VA Client*. Moreover, binning significantly affects the formation of information hierarchies (or datasets organized in tree-like fashion) that may influence the user's analytical and reasoning process. Although spatiotemporal data are by and large inherently hierarchical datasets, this is not the case for many others. Binning can effectively assist in the generation (or re-regeneration) of datasets initially featuring no explicit or simply flat structure. We should point out that in our case (re-)generating hierarchies from flat datasets is as vital as feature-engineering is in machine learning [5].

- **Visualization Chunks** are used as the internal unit of information exchanged between the different `Argus-Panoptes` components. A chunk is a `json` file containing the aggregated data for a given dataset hierarchy and the respective dataset schema. Over time, the schema may apparently change. To this end, the *VA Client* receives a visualization chunk once a request has been launched. The outcome of the `Spark` processing is a chunk and its main characteristics may either help improve or adversely impair system performance. Visualization chunks are tracked by the *VA Server* and in this respect, their invalidation, if needed, has to be an explicit user action.

We should point out that the process of (re-)generating features hierarchies in datasets is linearly correlated to both number and size of the produced visualization chunks. In this respect, we have established that in our experimentation discussed in this paper, the size of the largest size chunk generated is 142 MB and features $670K$ of data-rows. This chunk maintains the highest possible resolution and visualizes all features of the dataset.

### 4.1   The Architecture

Figure 3 outlines the architecture of our *IaaS*–based system: the server side is hosted on virtual computing systems as the left half of the figure shows, while the *VA Client* functionalities are shown to the right hand side. At the core of the server layout, the Spark-engine is referenced as a single entity although it may consist of a whole cloud cluster. It may also involve auxiliary services from the Hadoop ecosystem. In a minimal configuration, the computing cloud consists of an Apache Spark Master service deployed in standalone mode. A more common configuration would involve a Spark Master node, a number of Spark workers and a compatible distributed file system for storing and retrieving data such as HDFS. This example configuration could be extended with the addition of Apache Zookeeper for attaining high-availability as well as YARN or Mesos for cluster resource management.

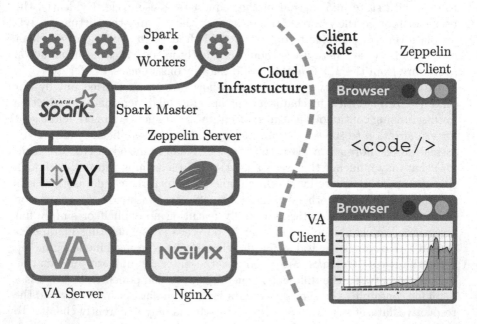

**Fig. 3.** Argus-Panoptes architecture layout

As Spark is not designed to offer a *REST API*, its connectivity with both Zeppelin and *VA Server* presents a point mismatched interface. Natively, Spark may only receive `.jar`, `.py` or .r jobs for Big-Data processing over the network. Here, our main concern is to maintain a single *SparkSession* at all times so that multiple clients dispatching jobs to the same visibility scope can be accommodated. Thus, continuity and on-line fashion processing for the client can be warranted. Apache Livy addresses the above issue as it can both provide a *SparkSession* and receive network-requests on behalf on the engine via its *REST API*. In this way, code submitted to Livy can be executed in the same visibility scope. For instance, if a user launches the code $\boxed{\texttt{val a="hello"}}$ with the help of a Zeppelin server connected to a Livy session, she can subsequently perform an *HTTP POST* request to Livy via the curl command-line tool and obtain the value of variable a. In our architecture, both *VA Server* and Apache Zeppelin are connected to Livy accessing the same *SparkSession*.

The Apache Zeppelin helps us realize the notion of interactive programming in the context of `Argus-Panoptes`. It is a platform that through its *paragraphs* allows for channelling tasks. On the left side of Fig. 2, we show the paragraphs as well as the controls of Zeppelin. This dialog-based portion of the Zeppelin-UI panel is the main interface facility for users to access the system.

Found in front of Livy, the *VA Server* carries a number of tasks and plays a central role for the coordinated operation of `Argus-Panoptes`. More specifically, the *VA Server*: *(1)* serves the *VA Client* with the JavaScript Web application. *(2)* dispatches code-segments for execution to Apache Spark. *(3)* receives visualization chunks from the Apache Spark. The latter are the outcome of Big-Data jobs executed at the engine. *(4)* maintains two-way communication channels with the *VA Client* with the help of WebSockets. *(5)* monitors the status of the *IaaS* computing resources and sends pertinent information update snippets to *VA Client* over time. *(6)* tracks visualization chunks produced so far and if requested explicitly by the user, it does carry out cache invalidation. *(7)* manages chunk-related information and offers an interface for profiling purposes. Developed with Ruby on Rails, the *VA Server* remains at all times "agnostic" in terms of specific characteristics of datasets under examination.

NginX is a reverse-proxy placed between our cloud-based components and *VA Client* (Fig. 3). The proxy is an additional layer for control and abstraction of resources/services and warrants smoother traffic flow between the interconnected servers and clients. In this manner, NginX has an invisible but crucial role as it effectively minimizes network traffic and consequently, enhances the perceived responsiveness of our platform. The main role of NginX is to forward *VA Client* chunk-requests to our server and let the client receive corresponding json files through *HTTP*. NginX transparently intercepts all outgoing json files and dispatches their gzip–ped versions. This leads to a non obvious performance improvement: The decompression of json files is handled automatically by a browser thread separate from the one that the JavaScript application is running thus the UI responsiveness is not halted during the decompression process.

## 4.2   The *VA Client* Functionality

Our *VA Client* is a `JavaScript` Web application that produces the entire visualization output interface. In this context, the `React/Redux` frameworks have been heavily used as they both promote component reusability and failure resistance. Figure 4 shows the output window of the UI after two operations have been requested: a drill-down for displaying crime in the London region and enhancement of the date dimension from quarterly to monthly.

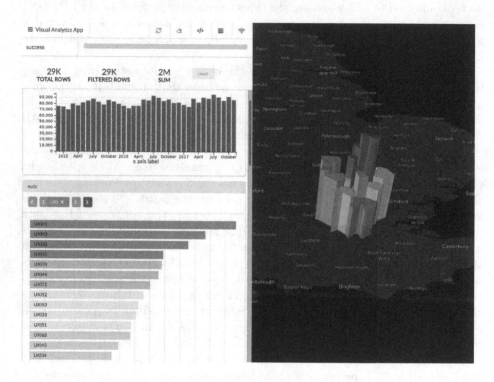

**Fig. 4.** *VA client* UI after a drill-down operation.

The *VA Client* uses the `Redux` framework as a mechanism of managing the local application state. `Redux` helps the application become independent from prior states. Similarly to the functional programming paradigm, the application interface generated at any point in time given a specific state, is always the same. As the schema of the data to be visualized next cannot be predicted, the interface cannot be constructed using information from the current state. We predominantly use the `React` framework for componentization. In a visual analytics application that caters for sophisticated users, the UI is an essential part of the architecture and invariably calls for much customization so that a application is both useful and timely. Hence, the one-fits-all solution approach is infeasible here. By offering components that can be readily reconfigured and reused, `React` plays a vital role in helping us put together effective UIs. In Fig. 4, every depicted visual element is a `React` component. Among others, there are components that deal with server connectivity, initiate cache invalidation, and refresh visualizations. There are also `React Bar/Row` *chart* components that help synthesize complex chart dashboards. The portion of UI to the right of Fig. 4 depicts map-based information and is constructed using a third-party `React Map` component [17].

## 5  Assessment with a Government *ASB* Dataset

Salient `Argus-Panoptes` features evolved during the prototype development. Experimentation with different real-world datasets from various disciplines also contributed to the realization of the system. In general, dataset features entail: *(1)* sized textual data, *(2)* raw tuple-based data for each incident that has received no aggregation, *(3)* geo-spatial features, *(4)* temporal features, and *(5)* other continuous or discrete features. In this section, we briefly present our experience with a publicly available dataset about crime. We use `Argus-Panoptes` as a spatial decision support analysis tool. Below, we discuss the pre-processing, the (re-)generation of feature hierarchies and our profiling of `Argus-Panoptes`.

The utilized dataset is curated and published by *UK Home Office* [9]. It maintains individual crime and anti-social behavior (ASB) incidents including street-level location information and is published in `CSV` format. All features in the dataset, are textual with *longitude* and *latitude* being numeric. Table 1 shows all features among with the number of distinct and null values for each feature.

**Table 1.** UK AST dataset key characteristics

| Name | Distinct | Null | Description |
|------|----------|------|-------------|
| crime_id | 12665725 | 5510538 | Incident identifier string |
| longitude | 737765 | 301155 | Longitude |
| latitude | 731070 | 301155 | Latitude |
| location | 280694 | 0 | Human-readable approximate location |
| lsoa_code | 35921 | 778773 | UK-designated area code |
| lsoa_name | 35065 | 778773 | UK-designated area name |
| reported_by | 46 | 0 | Reporting department |
| falls_within | 46 | 0 | Department with jurisdiction |
| month | 35 | 0 | Date string formatted as %Y-%m |
| last_outcome_category | 26 | 5806479 | Last outcome category |
| crime_type | 14 | 0 | Category of crime |
| context | 0 | 18268085 | Deprecated field |

In our pre-processing phase, we transform and store the dataset in a format suitable for our analyses. In particular, both Spark Master and Spark Worker nodes should be able to import the format in question correctly. For the dataset of Table 1, we carry out the following preprocessing steps: *(1)* transform date found in the *month* field to Date datatype, *(2)* drop the deprecated field *context* as well as the *crime_id*. *(3)* drop fields *lsoa_code*, *lsoa_name* and *location* deemed as redundant information, *(4)* save the DataFrame in an efficient columnar data representation, namely Apache Parquet.

We also transform the UK dataset by joining it with the NUTS classification scheme of Eurostat [6]. This enhancement offers varying granularity in regional information that has the following 4 levels: country (NUTS_0), major socio-economic region (NUTS_1), basic region (NUTS_2), and small region (NUTS_3). We use the Magellan [14] Spark library to perform the geo-join between the coordinates of each point and the *area polygon* of each region of the NUTS scheme. The geo-join helps us obtain the NUTS dimension which has only 178 distinct values, whereas the distinct values of the prior coordinate features were $760K$ (Fig. 6). This geo-join operation is CPU-intensive but it occurs only once and so, we make the data persistent for further processing.

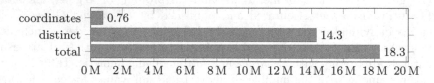

**Fig. 5.** Dataset has 18.3M rows: 14.3M distinct rows and 760K distinct coordinates.

We also tinker with two more dimensions: *crime_type* and *date*. Through binning, we create 3 distinct types of crime: theft-related, anti-social behavior, and others. Then, we map the original 14 crime types to populate the 3 new bins. Similarly, we bin the time attribute of the dataset to populate quarterly and yearly levels. Figure 6 reveals distinct counts for all features of the dataset after introducing hierarchies. In contrast to the geo-joining, binning is an inexpensive operation and can be carried out on-line without affecting the system responsiveness. The latter is highly desirable as it affords the user to instantly experiment with the introduced hierarchies on-line and if needed, realign them on the spot.

The aforementioned generation of hierarchical dimensions results to a maximum of 24 distinct chunks. Figure 7 shows the computation time required for each of these chunks in conjunction with the number of visualization tuples each one contains. It takes anywhere between 7.10 and 16.80 s for chunks to be computed. The above range represents an acceptable delay as the computation of each chunk occurs only once. Through caching, subsequent accesses to already computed chunks is only dependent to the volume of the data ultimately transported over the network to the client.

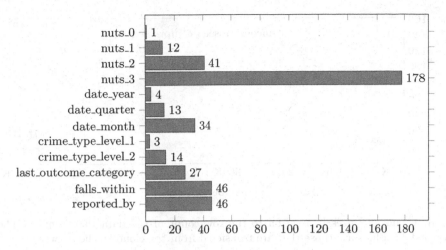

**Fig. 6.** Distinct counts for every feature resulting from geo-joining and data binning.

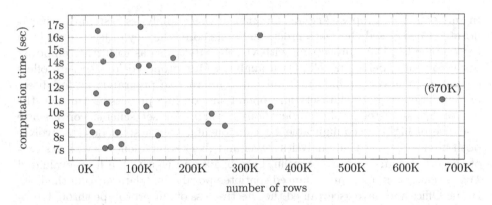

**Fig. 7.** Visualization chunk computation times in relation to the row count of the aggregated data. Big-Data aggregation operations are taking several seconds to complete and are independent of amount of data returned.

Figure 8 depicts the `json` and the corresponding compressed file sizes for all 24 different types of chunks. If `json` files were transported uncompressed, their size would range from 1.56 MBytes up to 140.90 MBytes. In actuality, all such files are transfered `gzip`-ped and their sizes ranges between 0.08 MBytes and 8.70 MBytes with average chunk size being less than 2.00 MBytes. Such sizes facilitate both the sought on-line type of operation and accomplish responsiveness for our prototype. Last but not least, we should indicate that a large number of visual interface interactions can be immediately served by already cached content in *VA Client*.

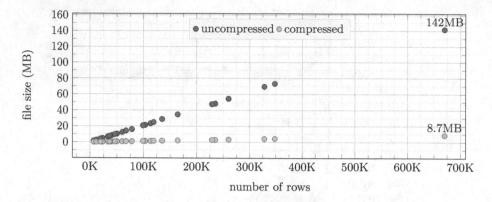

**Fig. 8.** Chunk file sizes in relation to the row count of the data they contain. The gzip-ped json files are those that are transfered from the cloud to the browser.

## 6    Concluding Remarks

In this paper, we propose Argus-Panoptes, a visual analytics system that incorporates cloud-based Big-Data processing in its core. Our key objective has been to combine Big-Data processing with visual analytics so as to further empower both domain experts and Big-Data analysts. Our proposed architecture offers a number of novel mechanisms that entail interactive programming for direct manipulation of both datasets and operations, on-line processing through the use of Spark-clusters, robust operations through dataset schema convergence and use of highly reconfigurable UI components. Our system design involves both home-grown virtual analytics server and client components as well as state-of-the-art systems such as Zeppelin, Livy, Spark and NginX. We have evaluated Argus-Panoptes using an enhanced spatiotemporal crime dataset from the U.K. Home Office and have ascertained the effectiveness of our prototype through profiling of its operations.

## References

1. Apache Zeppelin: Zeppelin: web-based notebook (2009). https://zeppelin.apache.org. Accessed 30 June 2018
2. Cloudera: Hue is an open source analytics workbench for self service BI. (2009). http://gethue.com. Accessed 30 June 2018
3. Daniel, K., Kohlhammer, J., Ellis, G., Mansman, F. (eds.): Mastering the Information Age Solving Problems with Visual Analytics. Eurographics Association (2010)
4. Dibia, V., Demiralp, Ç.: Data2Vis: automatic generation of data visualizations using sequence to sequence recurrent neural networks, April 2018. arxiv.org/abs/1804.03126
5. Domingos, P.: A few useful things to know about machine learning. Commun. ACM **55**(10), 78–87 (2012)

6. EUROSTAT: NUTS - nomenclature of territorial units for statistics (2016). http://ec.europa.eu/eurostat/web/nuts/background. Accessed 30 June 2018
7. Facebook Inc.: React: a JavaScript library for building user interfaces (2009). https://reactjs.org. Accessed 30 June 2018
8. Fekete, J.D.: Visual analytics infrastructures: from data management to exploration. Computer **46**(7), 22–29 (2013)
9. Home Office, UK: ASB incidents, crime and outcomes (2015). https://data.police.uk/about/. Accessed 30 June 2018
10. Jupyter Team: Jupyter project (2009). https://jupyter.org. Accessed 30 June 2018
11. Keim, D.A.: Visual exploration of large data sets. Commun. ACM **44**(8), 38–44 (2001)
12. Liu, Z., Jiang, B., Heer, J.: ImMens: real-time visual querying of Big Data. Comput. Graph. Forum **32**(3), 421–430 (2013)
13. Novus Partners: NVD3: reusable charts for d3.js (2014). http://nvd3.org. Accessed 30 June 2018
14. Sriharsha, R.: Magellan: geospatial analytics using spark (2015). https://github.com/harsha2010/magellan. Accessed 30 June 2018
15. Siddiqui, T., Kim, A., Lee, J., Karahalios, K., Parameswaran, A.: Effortless data exploration with zenvisage: an expressive and interactive visual analytics system. Proc. VLDB Endow. **10**(4), 457–468 (2016)
16. Thomas, J.J., Cook, K.A.: Illuminating the path: the research and development agenda for visual analytics. IEEE Computer Society (2005). http://vis.pnnl.gov/pdf/RD_Agenda_VisualAnalytics.pdf
17. Uber: Deck.gl large-scale WebGL-powered data visualization. https://uber.github.io/deck.gl
18. Vartak, M., Huang, S., Siddiqui, T., Madden, S., Parameswaran, A.: Towards visualization recommendation systems. ACM SIGMOD Rec. **45**(4), 34–39 (2017)
19. Wong, P.C., Shen, H.W., Johnson, C.R., Chen, C., Ross, R.B.: The top 10 challenges in extreme-scale visual analytics. IEEE Comput. Graphics Appl. **32**(4), 63–67 (2012)
20. Wongsuphasawat, K., et al.: Voyager 2. In: Proceedings of 2017 CHI Conference on Human Factors in Computing Systems (CHI 2017), Denver, pp. 2648–2659, May 2017)
21. Zaharia, M., et al.: Resilient distributed datasets: a fault-tolerant abstraction for in-memory cluster computing. In: Proceedings of 9th USENIX Conference on Networked Systems Design and Implementation (NSDI 2012), San Jose (2012)

# An Overview of Big Data Issues
# in Privacy-Preserving Record Linkage

Dinusha Vatsalan[1], Dimitrios Karapiperis[2(✉)], and Aris Gkoulalas-Divanis[3]

[1] Data61, CSIRO, Sydney, Australia
`dinusha.vatsalan@data61.csiro.au`
[2] School of Science and Technology, Hellenic Open University, Patras, Greece
`dkarapiperis@eap.gr`
[3] IBM Watson Health, Cambridge, MA, USA
`gkoulala@us.ibm.com`

**Abstract.** Nearly 90% of today's data have been produced only in the last two years! These data come from a multitude of human activities, including social networking sites, mobile phone applications, electronic medical records systems, e-commerce sites, etc. Integrating and analyzing this wealth and volume of data offers remarkable opportunities in sectors that are of high interest to businesses, governments, and academia. Given that the majority of the data are proprietary and may contain personal or business sensitive information, Privacy-Preserving Record Linkage (PPRL) techniques are essential to perform data integration. In this paper, we review existing work in PPRL, focusing on the computational aspect of the proposed algorithms, which is crucial when dealing with Big data. We propose an analysis tool for the computational aspects of PPRL, and characterize existing PPRL techniques along five dimensions. Based on our analysis, we identify research gaps in current literature and promising directions for future work.

**Keywords:** Privacy-Preserving Record Linkage · Entity resolution

## 1 Introduction

In the era of information explosion, massive amounts of data, coming from various sources, need to be integrated to facilitate data analysis for businesses, governments, and academia. Record linkage, also known as *entity resolution* or *data matching*, is the process of resolving whether two records that belong to disparate data sets, refer to the same real-world entity. Record linkage is a two-step process. The goal of the first step, known as *blocking*, is to formulate as many as possible matching pairs and, simultaneously, maintain the number of non-matching pairs as small as possible. In the second step, termed as *matching*, the distances between the pairs formed during the blocking step are calculated. Privacy-Preserving Record Linkage (PPRL) investigates how to perform the steps described above in a secure manner, by respecting the privacy of the individuals who are represented in the data. For this reason, input records undergo

© Springer Nature Switzerland AG 2019
Y. Disser and V. S. Verykios (Eds.): ALGOCLOUD 2018, LNCS 11409, pp. 118–136, 2019.
https://doi.org/10.1007/978-3-030-19759-9_8

a data masking process that embeds them into a space, where the underlying data is kept private (Fig. 1).

In this paper, we adapt the taxonomy proposed for PPRL in [69] and – inspired from analysis tools that are commonly used in business – such as SWOT and PEST [23,54], we develop an analysis tool that focuses on the computational aspects of PPRL techniques. We describe the proposed analysis tool for the computational aspects of PPRL in Sect. 2. In Sect. 3, we use this tool to characterize, analyze, review and compare existing PPRL techniques with respect to their computational aspects. Last, in Sect. 4, we discuss a number of gaps that we identified in current literature, along with some promising directions for future research.

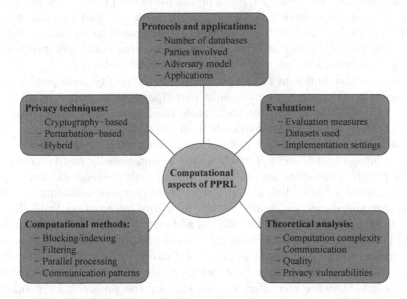

**Fig. 1.** An analysis tool consisting of five dimensions used to analyze and characterize computational aspects of PPRL techniques.

## 2    Analysis Tool

In this section, we present our proposed analysis tool for the computational aspects of PPRL techniques. It consists of five dimensions, each of which includes several topics. These dimensions are: (1) protocols and applications, (2) privacy techniques, (3) computational methods, (4) theoretical analysis, and (5) evaluation. In what follows, we describe each of these dimensions in detail.

### 2.1    Protocols and Applications

This dimension includes the protocol settings and application areas of PPRL techniques. The PPRL protocol is determined by the number of databases to be linked, the parties involved, and the considered adversarial model. The application areas of PPRL specify the different computation aspects, such as volume, flow, real-time vs batch processing, dynamic nature, and sensitivity of errors and variations in the data.

**Number of databases.** The computational aspect of PPRL is associated with the number of databases that have to be linked using the PPRL protocol. The naïve comparison space required for PPRL has an exponential growth with the number of databases. Existing techniques for PPRL can be categorized into *two databases linking* and *multiple databases linking*, where the latter received a lot of attention recently, due to the increasing demand of supporting Big Data applications [74, 76]. The computational challenges and privacy risk in terms of collusion between parties, with the aim to learn another party's data, increase with the number of databases to be linked. Further, the variations and different schemas used in different databases, result in the linkage quality challenge, which requires advanced techniques.

**Parties involved.** Different PPRL protocols use different types of parties for the linkage. The database owners typically participate in the protocol with the use of an external linkage unit for facilitating the linkage. The linkage unit conducts linkage of the encoded records from the database owners. Some PPRL protocols use more than one linkage unit, which leads to additional privacy risks. Linkage unit-based approaches are computationally more efficient that other PPRL approaches, especially for linking multiple large databases, since the protocols without linkage unit need more complex techniques to make sure that the database owners cannot infer any information from the data that is exchanged among them [69]. In addition, a global authority might be used in some protocols for managing or providing encoding keys and protocol parameters to the parties of the protocol. Finally, a researcher or an external party may be involved in the protocol to obtain access to some attributes of the records that are identified as matching by the protocol, for conducting further analysis.

**Adversary model.** PPRL protocols generally assume either the *honest-but-curious model* (HBC), or the *malicious model* [21, 22, 46]. In the HBC model, parties are curious in that they try to find out as much as they can about the other party's inputs, through inference attacks or collusions, while following the protocol [22, 46]. Inference attacks can be performed on encoded records based on some background information, such as frequency distribution, to re-identify the records. Collusion is a privacy risk of some parties colluding among them to learn other parties' sensitive information [46]. The protocol is secure in the HBC perspective if and only if all parties involved obtain no new knowledge at the end of the protocol, above what they would have learned from the output. In contrast to HBC parties, in the malicious model the parties behave arbitrarily in terms of refusing to participate in a protocol, not following the protocol, choosing arbitrary values for their data inputs, or aborting the protocol at any time [45]. PPRL techniques under the malicious model are computationally expensive and privacy evaluation of PPRL techniques under this model is difficult compared to the HBC model, due to many potentially unpredictable ways of malicious parties to deviate from the specified steps of the protocol [9,21,46]. Since PPRL techniques for HBC models are not realistic for real-world applications and PPRL for malicious models are computationally expensive, more advanced models have been recently proposed in cryptography [46]. Two of them are the *accountable computing* and the *covert model*, where the former allows honest parties to detect the misbehavior of an adversary with high probability [2], and the latter provides accountability for privacy compromises without the excessive complexity and cost of the malicious model [27].

**Applications.** PPRL is increasingly being required in several application areas, including in healthcare, national security, crime and fraud detection, business, governments, social sciences and population informatics. For example, data from several sources including different hospitals, pharmacies, and travel data, need to be linked for outbreak detection or clinical trials in healthcare applications [8,48], while linking the social security databases, law enforcement agencies databases, the police databases, and Internet service providers databases allows identifying crimes and frauds in national security and crime and fraud detection applications [28,55,78]. The computational requirements of PPRL techniques depend on the application area they are developed for, including the type and size of data, required output, and other application-specific constraints.

## 2.2 Privacy Techniques

The privacy techniques used for encoding data, processing data, and comparing and classifying data in PPRL can be categorized as follows:

**Cryptographic-based techniques.** These employ computationally expensive secure multi-party computation (SMC) techniques, such as homomorphic encryptions, Yao-based protocols, secret sharing, secure scalar product, and secure vector operations [46]. Although these techniques are provably secure and highly accurate, they are not efficient and scalable enough to be used for linking large databases.

**Perturbation-based techniques.** These are computationally efficient methods, allowing PPRL to scale to large databases. Increasing the privacy of perturbation-based techniques, however, results in accuracy loss, and vice versa. Some of the widely used perturbation techniques include generalization (such as $k$-anonymity, value generalization hierarchies, and binning), noise addition techniques (such as random and differential privacy), embedding techniques, and probabilistic data structure-based approaches (such as Bloom filters and count-min sketches). Perturbation-based techniques, especially Bloom filters, have increasingly been used in several real PPRL applications in recent times [60].

**Hybrid PPRL approaches.** These use perturbation-based techniques to efficiently remove highly non-matching records from the comparison space and then apply cryptographic-based techniques on the resulting records to achieve high quality linkage results without excessive computation [26].

### 2.3   Computational Methods

Several computational methods have been proposed in the literature aiming to reduce the exponential comparison (or search) space required by naïve PPRL and to speedup the linkage process. These methods of optimization are largely orthogonal, so that they can be combined to achieve maximal efficiency.

**Blocking approaches.** Blocking is defined on selected attributes to partition the records in a database into several blocks, such that comparison can be restricted to the records of the same block. Numerous blocking techniques have been used for record linkage [11] and for PPRL, with the additional challenge in the case of PPRL of preserving privacy in the blocking step [69]. Blocking improves the runtime of linkage, but it still involves unnecessary comparisons that limit its performance. Block processing is the approach of restructuring a collection of generated blocks to be compared and classified in the next step, so that unnecessary comparisons are pruned [32,59].

**Filtering approaches.** Filtering is an optimization for a particular comparison function which optimizes the evaluation of a specific similarity measure for a predefined similarity threshold to be met by matching records. It utilizes filtering or indexing techniques to eliminate sets of records that cannot meet the similarity threshold for the selected similarity measures [10,16].

**Parallel processing approaches.** Parallel linkage aims at improving the execution time proportionally to the number of processors [15,40,41]. This can be achieved by partitioning the entire set of record pairs to be compared, and conducting the comparison of the different partitions in parallel on different processors. A special case would be to utilize a blocking approach to compare the records in different blocks in parallel. Two approaches have been used so far for parallel linkage: (1) utilizing graphics processing units (GPUs) [19,63], and (2) using Hadoop and its MapReduce framework [37,42,77].

**Communication patterns.** Different communication patterns have different computation and communication complexities. With the increasing number of databases, the comparison space remains very large, even when a blocking or indexing technique is used [56,72,74]. Improved communication patterns can reduce the exponential growth (for larger number of databases to be linked) down to a smaller value. Such improved communication patterns include sequential, ring-by-ring, tree-based, and hierarchical patterns. Some of these patterns have been recently used for PPRL on multiple databases [74].

### 2.4   Theoretical Analysis

The dimension of theoretical analysis of PPRL techniques includes analysis of complexity, quality, and privacy vulnerabilities to allow for comparison and assessment of their expected scalability to large databases, quality of linkage results, and privacy guarantees.

**Computation and communication complexity.** The overall computational efforts and cost of communication required in the PPRL process are generally measured using the big $O$-notation [53]. For example, given that $n$ is the number of records in a database, $O(n)$ represents linear complexity, $O(n^2)$ quadratic complexity, and $O(c^n)$ exponential complexity, where $c > 1$.

**Quality of linkage.** The quality of linkage is theoretically analyzed in terms of fault-tolerance of the linkage technique to data errors and variations, whether the matching records are identified across all databases or it allows identifying matching records across subsets of databases, trade-offs with privacy and complexity, step-wise quality (i.e., preprocessing quality, blocking quality, and matching quality), and final linkage quality.

**Privacy vulnerabilities.** The privacy vulnerabilities that a PPRL technique is susceptible to, provide a theoretical estimate of the privacy guarantees of the technique. These include *frequency attacks*, where the list of encoded values is matched with the frequency distribution of a list of unencoded values, *dictionary attacks*, where a list of unencoded values are matched with the list of encoded values by applying different encoding functions on the unencoded values, and encoding-specific attacks, such as *cryptanalysis attacks* specific to Bloom filter encoding, where – depending upon the parameter setting – iterative mapping of individual encoded values back to their original values is possible using a constrained satisfaction solver. Another vulnerability associated with linkage unit-based approaches and/or multiple-databases linking is *collusion* between parties, where parties involved in a PPRL protocol may work together to find out another party's data.

## 2.5  Evaluation

The linkage outcomes need to be evaluated in terms of scalability, linkage quality, and privacy. This dimension includes evaluation measures, datasets, and implementation settings.

**Evaluation measures.** The scalability and linkage quality can be evaluated using standard evaluation measures, such as runtime, memory consumption, communication size, speedup, reduction ratio, pairs completeness, pairs quality, precision, recall, and the F-measure [10,69,73]. However, linkage quality evaluation requires access to truth data, which can be rarely accommodated in PPRL applications. Consequently, sample evaluation or evaluation on synthetic/perturbed datasets is typically used to assess linkage quality. Various measures have been used to quantify the privacy protection of PPRL techniques, including information theory-based entropy and information gain measures [17,31,64], as well as disclosure risk-based measures [18,25,65,70,73]. However, no standard measures for privacy evaluation have been used in the literature.

**Datasets.** Experimental evaluation of PPRL techniques on several datasets is important to gain reliable evidence of the techniques' performance. Due to the difficulties of obtaining real-world data that contain personal information, synthetically generated or perturbed databases are typically used. Several tools are available to generate or corrupt data [13,24,66]. However, to evaluate PPRL techniques with regard to their expected performance in real-world applications, evaluations should ideally be done on databases that exhibit real-world properties and error characteristics.

**Implementation.** The implementation techniques that have been used to prototype a PPRL technique and the settings used for conducting its experimental evaluation, determine the complexity and scalability results. Further, some scalability measures, such as runtime, memory size, and communication size, are platform dependent. Comparing different techniques requires conducting experimental evaluation in the same platform and settings.

## 3   Literature Review

In this section, we review existing literature and categorize PPRL techniques along the dimensions of computational methods, which include: (1) blocking/indexing techniques, (2) block processing techniques, (3) filtering techniques, (4) parallel processing, and (5) improved communication patterns.

### 3.1   Blocking Techniques

Numerous blocking strategies [10] have been developed for record linkage and PPRL. Standard blocking groups records according to blocking criteria (known as *blocking key*), to partition all records into disjoint blocks. The blocking key values (BKVs) are the values of a selected attribute (e.g., zipcode), or the result of a function on one or several attribute values (e.g., the concatenation of the first two letters of last name and year of birth). Other blocking approaches include sorted neighborhood that sorts records according to a sorting key and only compares neighboring records within a certain window, and canopy clustering that results in overlapping clusters [10]. Multi-pass blocking is utilized to improve recall, where records are blocked according to different blocking keys, at the cost of a larger number of comparisons. In the following, we review blocking approaches for PPRL along the dimensions of the proposed tool.

Al-Lawati et al. [1] proposed a secure blocking protocol for linking two databases with the use of a linkage unit that assumed a *HBC* adversary model. Token-based blocking was used to improve computation efficiency. The linkage unit matches the records based on the computed TF-IDF distances of the hash signatures, using the *Jaccard* coefficient. The proposed blocking approach consists of three methods: *simple blocking*, *record-aware blocking*, and *frugal third party blocking* [1]. Simple blocking arranges hash signatures in overlapping blocks where the similarity of a pair may be computed more than once if they are in more than one common block. Record-aware blocking solves this issue by using an identifier with every hash signature to indicate the record it belongs to. Frugal third party blocking, uses a secure set intersection (SSI) protocol to reduce the cost of transferring the whole databases to the third party, by first identifying the hash signatures that occur in both databases.

Inan et al. [26] proposed a hybrid approach for PPRL of two databases under the *HBC* adversary model, by combining efficient generalization and expensive cryptographic privacy techniques. A blocking approach based on value generalization hierarchies is used, and the record pairs that need more detailed comparison to determine the match status are compared in a computationally expensive *SMC* computation step, using cryptographic techniques. The cost is reduced in the blocking step by reducing the number of candidate record pairs that need to be compared using SMC techniques.

A secure blocking based on phonetic encoding algorithms was presented by Karakasidis et al. [29]. A *three-party* (two database owners and a linkage unit) setting in a *HBC* model is assumed. The basic idea is to encode the values of a BKV (e.g. last name) with a phonetic function, such as Soundex or Metaphone [10]. All records with the same phonetic code are assigned to the same block. This approach uses a secure version of *edit distance* on *Bloom filters*. The experimental study, conducted on a synthetic dataset (generated using *Febrl* [12]), showed that the approach outperforms the original edit distance algorithm in terms of complexity (due to the secure blocking component), while preserving privacy, and also offers almost the same matching performance.

Karakasidis et al. [31] proposed three noise addition techniques for improving the privacy of [29]. In the first method, fake values are added to the datasets, such that both the attribute values and the Soundex values exhibit uniform distributions. This increases the complexity due to excessively oversized datasets. The second method overcomes this drawback by modifying the frequency of attribute values, such that all Soundex values occur equally frequent. However, some attribute values are removed where the corresponding Soundex values have more than the average number of attribute values, and therefore true matches might be missed in the linkage process. The third method adds fake values, where each Soundex value reflects at least $k$ attribute values. Parameter $k$ is tunable to adjust the number of fake records added in order to balance the trade-off between complexity and privacy. This work was experimentally evaluated using a real Australian telephone database, and the results indicated that in terms of information gain, using a phonetic-based fake injection blocking approach can offer adequate privacy for PPRL.

A generalization-based $k$-anonymous private blocking approach using a reference table was proposed in [30] for linking two databases using a linkage unit. Initially, clusters of size $k$ are generated for the set of reference values that are shared by the database owners, using the $k$-nearest neighbor clustering algorithm with the Dice coefficient metric. Then, each database owner assigns their set of BKVs to the respective clusters. The resulting clusters are sent to the linkage unit that identifies and merges the corresponding clusters to generate candidate record pairs. Clusters contain at least $k$ reference values, making inference attacks using the reference values difficult. Experiments conducted using a real Australian telephone book as the reference table and synthetic data generated using *Febrl* [12], as datasets to be linked, validate that this approach provides $k$-anonymity guarantees, while reducing the number of candidate record pairs.

Vatsalan and Christen in [71] proposed an approach that utilizes local sorted neighborhood clustering for improved performance in the blocking phase, to generate $k$-anonymous clusters based on reference values. Each database owner sorts a shared set of reference values and then inserts their records into the sorted list according to their sorting keys. Initial Sorted Neighborhood Clusters are determined such that each cluster contains one reference value and a set of database records. To offer $k$-anonymity, the initial clusters are merged into larger blocks containing at least $k$ database records based on similarity or size constraints. These clusters are then sent to a linkage unit to identify and merge similar clusters across the databases, based on the reference values in the clusters. Experiments conducted on real Australian telephone and North Carolina voters databases, show the improved performance of the approach compared to [30], in terms of runtime and blocking quality. A variant of [71], involving a two-party setting without a linkage unit, was presented in [75]. In this approach, the two database owners generate their reference values independently. Each database owner sorts its reference values, inserts its records into the sorted list, builds initial clusters (with one reference value and its associated records), and merges these clusters to guarantee $k$-anonymity. Afterwards the database owners exchange their reference values, which are then merged together and sorted. In order to find candidate pairs between the sources, a sorted neighborhood method with a sliding window $w$ is applied on the reference values. The window size $w$ determines the number of reference values originating from each data source in a window. The clusters of the reference values that fall into the same window are determined as candidate blocks, which need to be compared in the next step. An evaluation study, conducted in [73] using real and synthetic datasets, showed that this approach outperforms several existing blocking approaches in terms of runtime and privacy, with no loss in blocking quality.

Durham investigated the use of Locality-Sensitive Hashing (LSH) [20] for private blocking of records encoded as Bloom filters [17]. She proposed the use of a family of hash functions (Minhash for Jaccard, or Hamming LSH for Hamming distances) to generate keys that are used to partition the records in a database, so that similar records are grouped into the same block [39]. A Minhash function permutes the bits in a Bloom filter and selects the first index position in the permuted Bloom filter that is set to 1. By applying $\phi$ Minhash functions, $\phi$ index positions are obtained, which are concatenated to generate the final Minhash key for the Bloom filter. The HLSH hash functions select the bit value of a Bloom filter at a random position. In the same way as Minhash, $\phi$ HLSH functions are applied on a record's Bloom filter and the values of the $\phi$ selected bits are concatenated to obtain the final hash key. LSH provides guaranteed accuracy while being efficient. However, it requires data dependent parameters to be tuned effectively and it can be applied only to specific encodings, such as Bloom filters.

Karapiperis et al. proposed a private blocking approach for linking multiple databases based on LSH [34–36]. This approach uses $L$ independent hash tables, each consisting of key-bucket pairs, where keys represent the blocking keys and buckets host a linked list aimed at grouping similar records. Each hash table is assigned a set of $K$ hash functions, generated by a linkage unit and sent to all the database owners to populate their set of blocks. This approach was later extended by proposing a frequent pairs scheme (FPS) [38] for reducing the number of comparisons, while maintaining a high level of recall. FPS achieves high blocking quality by identifying similar record pairs that exhibit many LSH collisions, and then performs distance calculations only for those pairs. Empirical results showed significant improvement in running time due to a drastic reduction of candidate pairs by FPS, while achieving high blocking quality [35]. Based on these methods, the authors have developed LSHDB [33], which uses LSH to efficiently block the masked records, and store the produced blocking structures on disk for further use. LSHDB achieves very fast response times, which makes it ideal for online settings, thanks to the utilization of efficient algorithms and the employment of flexible and robust noSQL systems for storing the data.

Ranbaduge et al. [56] proposed a private blocking approach for multiple databases without a linkage unit, based on a single-bit tree data structure. The single-bit tree is iteratively constructed by all database owners to store records' Bloom filters, such that similar records are placed into the same tree leaf. At each iteration, the set of Bloom filters in a tree node is recursively split based on selected bit positions, which are agreed upon by all parties. This requires a communication step among all parties in each iteration of the algorithm. Another drawback of this approach is that it might miss true matches due to the recursive splitting of Bloom filters. This limitation was addressed in [57], using a multi-bit tree [43] data structure combined with canopy clustering. Multi-bit trees were used to split the database records (encoded into Bloom filters) individually by the database owners into small mini-blocks, which are then merged into larger blocks according to privacy and computational requirements, using a canopy clustering technique [14].

A communication-efficient private blocking approach for multiple databases using a linkage unit, was proposed in [58]. In this approach, local blocks are generated individually by each database owner, using a private blocking technique. A block representative, in the form of a min-hash signature [7], is then generated for each block and sent to the linkage unit. The linkage unit applies global blocking using LSH to identify the candidate block sets from all databases, based on the similarity between block representatives. Local blocking enables the database owners to generate their blocks with more flexibility and control, without any iterative communication among them. This approach outperforms existing private blocking approaches for multiple databases in terms of scalability, privacy, and blocking quality [58].

## 3.2   Block Processing Techniques

Several block processing methods have been used for record linkage and PPRL, in order to process the generated blocks in an efficient and effective way to reduce the number of required comparisons, while improving blocking quality [51,52,59]. In this section we review these block processing techniques.

Wang et al. [79] introduced an iterative block processing technique for deduplicating a single database. The comparison results of blocks are propagated to subsequent blocks to avoid repeated comparisons. This approach was later extended in [39] for record linkage using LSH.

Two categories of block processing methods were used by Papadakis et al. [51] for deduplication. The first category includes block purging and block scheduling methods, which operate at the coarse level of processing individual blocks. The second category of comparison-refinement methods, such as comparison propagation, duplicate propagation, comparison pruning, and comparison scheduling, operate at a finer level of individual comparisons within blocks.

The concept of meta-blocking was introduced by Papakadis et al. in [52] for record linkage, to restructure a collection of blocks to reduce the number of comparisons. In their approach, a block collection is provided as input to a supervised classifier to identify promising comparisons based on block-feature vectors. The drawback of this approach is the selection of suitable features and requirement of training data to achieve accurate pruning of record comparisons.

Meta-blocking has been recently studied for the PPRL of two databases using a linkage unit [32]. A sorted neighborhood blocking based on reference values is used along with multi-sampling transitive closure as a processing technique to prune records based on redundant assignments to blocks. Experimental results show the efficacy of the approach in terms of recall and computational cost.

Recently, a general meta-blocking technique for PPRL on multiple databases was proposed by Ranbaduge et al. [59]. Their approach uses a graph structure to schedule the comparison of blocks with the aim of minimizing the number of repeated and superfluous comparisons between records, where the former is comparison of duplicate record pairs and latter is comparison of records with non-matching record pairs. The experimental results of their approach on real datasets, show that up to five orders of magnitude reduction in the number of record comparisons can be achieved compared to existing approaches.

## 3.3   Filtering Techniques

Several filtering approaches have been proposed in record linkage and PPRL literature to speed up the linkage process. The proposed optimizations, include the use of different filters, such as length and prefix filters, and dynamic inverted indexes [5]. Several filtering approaches also utilize the characteristics of similarity measures for metric spaces to reduce the search space, such as the triangle inequality [80]. For PPRL, filtering approaches need to be adapted to the comparison of encoded records, such as Bloom filters. In what follows, we describe several filtering approaches that have been proposed for PPRL.

Token-based similarity functions, such as the Jaccard and Dice coefficients, allow the application of a simple *length filter* to reduce the comparison space. This is because the minimal similarity can only be achieved if the lengths (for example, the number of bits set to 1 in Bloom filters) of the two records do not deviate too much. Formally, for records $r_i$, $r_j$, with $|r_i| \leq |r_j|$, it holds that: $Jacc\_sim(r_i, r_j) \geq s_t \Rightarrow |r_i| \geq \lceil s_t \cdot |r_j| \rceil$ and $Dice\_sim(r_i, r_j) \geq \frac{2min(|r_i|,|r_j|)\cdot(1-s_t)}{s_t}$.

For example, two records cannot satisfy a Jaccard similarity threshold $s_t = 0.8$ if their lengths differ by more than 20%, and a Dice similarity threshold $s_t = 0.8$ if the length difference is at least 50% of the length of the smaller record. Hence for a similarity threshold of 0.8, the length filter would prune record pairs that do not meet the length condition without comparing in detail.

Vatsalan and Christen used such a length filter for Dice coefficient similarity for Bloom filter-based PPRL in a two-party setting without using a linkage unit [67]. In their approach, certain bit positions (depending on privacy criteria) from the Bloom filters are iteratively exchanged between the two database owners to classify the pairs as matches, non-matches, and possible matches. For the possible matches in an iteration, more bits are revealed in the next iteration until the maximum number of bits are revealed, or all pairs are classified as matches or non-matches. A length filtering phase is used in addition to phonetic blocking to filter record pairs that are potentially non-matches, without revealing any bit positions for these records' Bloom filters.

The privacy-preserving version of PPJoin (called P4Join), proposed by Sehili et al. [63], utilizes three filters to reduce the Bloom filter-based comparison space: the length filter, a prefix filter, and a position filter. The *prefix filter* excludes Bloom filter pairs that have an insufficient overlap in the bit positions set to 1, in order to satisfy a predefined threshold, and this overlap test can be limited to only the prefix bit positions of the Bloom filters. The *position filter* of P4Join can avoid the comparison of two records even if their prefixes overlap, depending on the prefix positions where the overlap occurs. However, these filtering approaches achieve only a small improvement for PPRL, since the filter tests incur significant cost. Moreover, Bloom filter encoding for PPRL should ideally have 50% of their bits set to 1, in order to make them less vulnerable to frequency attacks [49], thereby constituting filtering less effective.

The use of multi-bit trees was proposed for fast similarity search in large databases of chemical *fingerprints* (encoded into Bloom filters) [3,43]. A multi-bit tree is a binary tree used to iteratively assign fingerprints to its nodes based on match bits. A match bit refers to a specific position of the bit vector, and can be 1 or 0: it indicates that all fingerprints in the associated subtree share the specified match bit. When building up the multi-bit tree, one match bit, or multiple such bits, are selected in each step, so that the number of unassigned fingerprints can be roughly split in half. The split is continued as long as the number of fingerprints per node does not fall under a limit. The match bits can then be used for a query fingerprint, to determine the maximal possible similarity for subtrees when traversing the tree and can thereby eliminate many fingerprints to compare. The multi-bit tree-based approach was extended by Bachteler et al. [3] to partition the fingerprints according to their lengths, such that all fingerprints with the same length belong to the same partition. To apply the length filter, the search for similar fingerprints using a Jaccard similarity measure is restricted to the partitions meeting the length criterion of $Jacc\_sim(r_i, r_j) \geq s_t$. Query efficiency is further improved by organizing all fingerprints of a partition within a multi-bit tree. Experimental evaluations showed that the multi-bit tree approach is very effective and performs equally or superior to blocking approaches, such as canopy clustering and sorted neighborhood [3,62].

Several metric space-based PPRL approaches have been proposed in the literature. One of the main properties that a metric or distance function for metric spaces has to satisfy is the triangle inequality. Distance functions for metric spaces satisfying this property include the Euclidean distance, edit distance, Hamming distance and Jaccard coefficient (but not Dice coefficient) [80]. The triangle inequality has been used for private comparison and classification in PPRL, using reference values [50,68], as well as a filtering technique to reduce the comparison space for similarity search and record linkage [4,6]. The triangle inequality allows to eliminate the computation of distance between two objects, based on their distances to a reference object or pivot.

### 3.4  Parallel Processing

The utilization of GPUs that provide thousands of cores within a single machine to speed-up similarity computations is a relatively new approach for parallel processing [19]. At the same time, Hadoop provides programming frameworks, such as MapReduce, Spark, and Flink, that allow developing programs to be automatically executed in parallel on Hadoop clusters [42,77]. In the following, we review two existing parallel PPRL techniques based on these two approaches.

A GPU-based parallel PPRL approach using the P4Join filtering is described in [63]. The approach sorts the records encoded into Bloom filters according to the number of bits set to 1, and partitions the set of Bloom filters into equi-sized blocks, such that multiple of such blocks fit into the GPU memory. Pairs of blocks are then continuously loaded into the GPU for parallel comparison. Length filtering and prefix filtering are applied to remove pairs of blocks that do not meet the filtering criterion, to reduce the the number of comparisons. Experimental evaluation results show that the approach improved runtime by a factor of 20, even with a low-profile graphics card (Nvidia GeForce GT 540M).

Several record linkage approaches have utilized the Hadoop-based MapReduce framework for parallel processing [42,77]. The Map tasks read the input data and assign each record to a block according to its blocking key value. Then, the records are redistributed among the Reduce tasks, such that all records with the same blocking key value are sent to the same Reduce task. Comparison is then performed in parallel by the Reduce tasks. The load balancing problem with highly skewed block sizes for parallel processing, is addressed in [42].

Karapiperis and Verykios proposed a parallel PPRL approach for linking two databases with a linkage unit using MapReduce [37]. The approach uses a

LSH-based blocking in the Map phase and determines the Minhash signature for each record encoded into a Bloom filter. These signatures are fragmented into several pieces and the Bloom filters are redistributed such that all Bloom filters with the same Minhash fragment value are assigned to the same Reduce task for comparison. The approach thus leads to a replicated redistribution of Bloom filters according to the number of fragments and a Bloom filter may have to be compared at several Reduce tasks. To overcome this problem, an alternative approach of chaining two MapReduce jobs was proposed, where the first job outputs the pairs of records' identifiers in the Reduce phase. In the second job, duplicate record pairs are grouped at the same Reducer to be compared only once. The evaluation of this approach in [37] shows the efficiency of parallel processing. However, the study was limited to a few nodes and only 300K records.

## 3.5 Improved Communication Patterns

Most PPRL techniques use the naïve all-to-one communication, where all database owners send their encoded records to a linkage unit to conduct the linkage. A few PPRL techniques for linking multiple databases use a ring communication, where encoded records are sent from one database owner to another, following a ring pattern [69]. Another communication pattern is all-to-all communication, where each database owner sends encoded records to all other database owners [44]. Last, some PPRL techniques require communication in several steps, or iteratively in many rounds [56], making them impractical for real applications.

Several query tree representations have been used for optimizing multi-way join queries [47,61], and can be adapted for efficient processing of multi-party PPRL. Schneider and DeWitt [61] studied query processing plans with different types of structures: left-deep, right-deep, and bushy. Left-deep and right-deep trees use a base table as the inner and outer operand, respectively, of each join in the plan, while in bushy trees both inputs to a join may themselves result from joins. For PPRL, the concepts used in deep trees can be adapted to improve efficiency. However, only one recent study investigated such improved communication patterns for linking multiple databases in PPRL.

Recent work by Vatsalan et al. [74] proposed two improved communication patterns for reducing the number of comparisons for PPRL on multiple databases using counting Bloom filters (CBFs). In [74], the parties are grouped into rings and a secure summation protocol is used to generate a CBF for each set of parties' records encoded into Bloom filters. The comparison of records is conducted: (1) sequentially by a $LU$, such that only the matches of a ring are compared with the candidate record sets of the next ring, or (2) symmetrically, without a $LU$, where matches are identified for each individual ring in the first phase and then, using the matches from individual rings, the matches from all rings are identified in the second phase. The computational complexity of MP-PPRL techniques is exponential in the number of records per database ($n^p$, assuming $n$ records in each of the $p$ databases). These improved communication patterns reduce this exponential growth with $p$ down to the ring size $r$ (with $r < p$).

# 4   Conclusions

Computational aspects of Privacy-preserving record linkage (PPRL) are crucial for making PPRL viable for linking large collections of disparate data sources, especially for Big data applications. In this paper we proposed an analysis tool for analyzing, reviewing, and comparing existing computational methods for PPRL and then conducted an extensive survey using the proposed tool. Such an analysis tool allows identifying research gaps in the current literature and promising directions for future work in PPRL.

# References

1. Al-Lawati, A., Lee, D., McDaniel, P.: Blocking-aware private record linkage. In: IQIS, pp. 59–68 (2005)
2. Aumann, Y., Lindell, Y.: Security against covert adversaries: efficient protocols for realistic adversaries. J. Cryptology **23**(2), 281–343 (2010)
3. Bachteler, T., Reiher, J., Schnell, R.: Similarity Filtering with Multibit Trees for Record Linkage. Tech. Rep. WP-GRLC-2013-01, German Record Linkage Center (2013)
4. Barros, J.E., French, J.C., Martin, W.N., Kelly, P.M., Cannon, T.M.: Using the triangle inequality to reduce the number of comparisons required for similarity-based retrieval. In: Electronic Imaging: Science & Technology, pp. 392–403 (1996)
5. Bayardo, R.J., Ma, Y., Srikant, R.: Scaling up all pairs similarity search. In: WWW, Canada, pp. 131–140 (2007)
6. Berman, A., Shapiro, L.G.: Selecting good keys for triangle-inequality-based pruning algorithms. In: IEEE Workshop on Content-Based Access of Image and Video Database, pp. 12–19 (1998)
7. Broder, A.Z.: On the resemblance and containment of documents. In: Compression and Complexity of Sequences, pp. 21–29. IEEE (1997)
8. Brook, E., Rosman, D., Holman, C.: Public good through data linkage: measuring research outputs from the western Australian data linkage system. Aust NZ J. Public Health **32**, 19–23 (2008)
9. Canetti, R.: Security and composition of multiparty cryptographic protocols. J. Cryptol. **13**(1), 143–202 (2000)
10. Christen, P.: Data Matching - Concepts and Techniques for Record Linkage, Entity Resolution, and Duplicate Detection. Data-Centric Systems and Application. Springer, Heidelberg (2012). https://doi.org/10.1007/978-3-642-31164-2
11. Christen, P.: A survey of indexing techniques for scalable record linkage and deduplication. IEEE TKDE **24**(9), 1537–1555 (2012)
12. Christen, P., Gayler, R., Hawking, D.: Similarity-aware indexing for real-time entity resolution. In: ACM CIKM, Hong Kong, pp. 1565–1568 (2009)
13. Christen, P., Pudjijono, A.: Accurate synthetic generation of realistic personal information. In: Theeramunkong, T., Kijsirikul, B., Cercone, N., Ho, T.-B. (eds.) PAKDD 2009. LNCS (LNAI), vol. 5476, pp. 507–514. Springer, Heidelberg (2009). https://doi.org/10.1007/978-3-642-01307-2_47
14. Cohen, W.W., Richman, J.: Learning to match and cluster large high-dimensional data sets for data integration. In: ACM SIGKDD, Edmonton, pp. 475–480 (2002)

15. Dal Bianco, G., Galante, R., Heuser, C.A.: A fast approach for parallel dedupli-
    cation on multicore processors. In: ACM Symposium on Applied Computing, pp.
    1027–1032 (2011)
16. Dey, D., Mookerjee, V., Liu, D.: Efficient techniques for online record linkage. IEEE
    Trans. Knowl. Data Engin. **23**(3), 373–387 (2010)
17. Durham, E.: A framework for accurate, efficient private record linkage. Ph.D. the-
    sis, Faculty of the Graduate School of Vanderbilt University, Nashville, TN (2012)
18. Elliot, M., Hundepool, A., Nordholt, E., Tambay, J., Wende, T.: Glossary on sta-
    tistical disclosure control. In: Joint UNECE/Eurostat Work Session on Statistical
    Data Confidentiality (2005)
19. Forchhammer, B., Papenbrock, T., Stening, T., Viehmeier, S., Draisbach, U.,
    Naumann, F.: Duplicate Detection on GPUs. In: Database Systems for Business,
    Technology, and Web, pp. 165–184 (2013)
20. Gionis, A., Indyk, P., Motwani, R.: Similarity search in high dimensions via hash-
    ing. In: VLDB, pp. 518–529 (1999)
21. Goldreich, O.: Foundations of Cryptography: Basic Applications, vol. 2. Cambridge
    University Press, Cambridge (2004)
22. Hall, R., Fienberg, S.: Privacy-preserving record linkage. In: PSD, Corfu, Greece,
    pp. 269–283 (2010)
23. Hill, T., Westbrook, R.: Swot analysis: it's time for a product recall. Long Range
    Plann. **30**(1), 46–52 (1997)
24. Hoag, J., Thompson, C.: A parallel general-purpose synthetic data generator. ACM
    SIGMOD **36**, 19–24 (2007)
25. Hundepool, A., et al.: Handbook on statistical disclosure control. A Network of
    Excellence in the European Statistical System in the field of Statistical Disclosure
    Control (2010)
26. Inan, A., Kantarcioglu, M., Bertino, E., Scannapieco, M.: A hybrid approach to
    private record linkage. In: IEEE ICDE, Cancun, Mexico, pp. 496–505 (2008)
27. Jiang, W., Clifton, C., Kantarcıoğlu, M.: Transforming semi-honest protocols to
    ensure accountability. Data Knowl. Eng. **65**(1), 57–74 (2008)
28. Jonas, J., Harper, J.: Effective counterterrorism and the limited role of predictive
    data mining. Policy Anal. **584** (2006)
29. Karakasidis, A., Verykios, V.S.: Secure blocking+secure matching = secure record
    linkage. JCSE **5**, 223–235 (2011)
30. Karakasidis, A., Verykios, V.S.: Reference table based k-anonymous private block-
    ing. In: ACM SAC, Riva del Garda, pp. 859–864 (2012)
31. Karakasidis, A., Verykios, V.S., Christen, P.: Fakling. In: Garcia-Alfaro, J.,
    Navarro-Arribas, G., Cuppens-Boulahia, N., de Capitani di Vimercati, S. (eds.)
    DPM/SETOP -2011. LNCS, vol. 7122, pp. 9–24. Springer, Heidelberg (2012).
    https://doi.org/10.1007/978-3-642-28879-1_2
32. Karakasidis, A., Koloniari, G., Verykios, V.S.: Scalable blocking for privacy pre-
    serving record linkage. In: Proceedings of the 21th ACM SIGKDD International
    Conference on Knowledge Discovery and Data Mining, pp. 527–536. ACM (2015)
33. Karapiperis, D., Gkoulalas-Divanis, A., Verykios, V.: LSHDB: a parallel and dis-
    tributed engine for record linkage and similarity search. In: ICDM Demo, pp. 1–4
    (2016)
34. Karapiperis, D., Gkoulalas-Divanis, A., Verykios, V.: Distance-aware encoding of
    numerical values for privacy-preserving record linkage. In: ICDE, pp. 135–138
    (2017)

35. Karapiperis, D., Verykios, V.: An LSH-based blocking approach with a homomorphic matching technique for privacy-preserving record linkage. TKDE **27**(4), 909–921 (2015)
36. Karapiperis, D., Verykios, V.: FEDERAL: a framework for distance-aware privacy-preserving record linkage. TKDE **30**(2), 292–304 (2018)
37. Karapiperis, D., Verykios, V.S.: A distributed framework for scaling up lsh-based computations in privacy preserving record linkage. In: ACM BCI, pp. 102–109 (2013)
38. Karapiperis, D., Verykios, V.S.: A fast and efficient hamming LSH-based scheme for accurate linkage. KAIS **49**(3), 1–24 (2016)
39. Kim, H., Lee, D.: Harra: fast iterative hashed record linkage for large-scale data collections. In: EDBT, Lausanne, Switzerland, pp. 525–536 (2010)
40. Kim, H., Lee, D.: Parallel linkage. In: ACM CIKM, pp. 283–292 (2007)
41. Kirsten, T., Kolb, L., Hartung, M., Groß, A., Köpcke, H., Rahm, E.: Data partitioning for parallel entity matching. VLDB **3**(2) (2010)
42. Kolb, L., Thor, A., Rahm, E.: Dedoop: efficient deduplication with hadoop. VLDB **5**(12), 1878–1881 (2012)
43. Kristensen, T.G., Nielsen, J., Pedersen, C.N.: A tree-based method for the rapid screening of chemical fingerprints. Algorithms Mol. Biol. **5**(1), 9 (2010)
44. Lai, P., Yiu, S., Chow, K., Chong, C., Hui, L.: An efficient Bloom filter based solution for multiparty private matching. In: International Conference on Security and Management, p. 7 (2006)
45. Lindell, Y., Pinkas, B.: An efficient protocol for secure two-party computation in the presence of malicious adversaries. In: Naor, M. (ed.) EUROCRYPT 2007. LNCS, vol. 4515, pp. 52–78. Springer, Heidelberg (2007). https://doi.org/10.1007/978-3-540-72540-4_4
46. Lindell, Y., Pinkas, B.: Secure multiparty computation for privacy-preserving data mining. JPC **1**(1) (2009)
47. Lu, H., Shan, M.C., Tan, K.L.: Optimization of multi-way join queries for parallel execution. In: VLDB, pp. 549–560 (1991)
48. Malin, B.A., El Emam, K., O'Keefe, C.M.: Biomedical data privacy: problems, perspectives, and recent advances. JAMIA **20**(1), 2–6 (2013)
49. Mitzenmacher, M., Upfal, E.: Probability and Computing: Randomized Algorithms and Probabilistic Analysis. Cambridge University Press, Cambridge (2005)
50. Pang, C., Gu, L., Hansen, D., Maeder, A.: Privacy-preserving fuzzy matching using a public reference table. In: McClean, S., Millard, P., El-Darzi, E., Nugent, C. (eds.) Intelligent Patient Management. Studies in Computational Intelligence, vol. 189, pp. 71–89. Springer, Heidelberg (2009).https://doi.org/10.1007/978-3-642-00179-6_5
51. Papadakis, G., Ioannou, E., Palpanas, T., Niederee, C., Nejdl, W.: A blocking framework for entity resolution in highly heterogeneous information spaces. IEEE Trans. Knowl. Data Eng. **25**(12), 2665–2682 (2013)
52. Papadakis, G., Papastefanatos, G., Koutrika, G.: Supervised meta-blocking. Proc. VLDB Endowment **7**(14), 1929–1940 (2014)
53. Papadimitriou, C.: Computational Complexity. Wiley, Hoboken (2003)
54. Peng, G.C.A., Nunes, M.B.: Using pest analysis as a tool for refining and focusing contexts for information systems research. In: Research Methodology for Business and Management Studies, Lisbon, Portugal, pp. 229–236 (2007)
55. Phua, C., Smith-Miles, K., Lee, V., Gayler, R.: Resilient identity crime detection. IEEE TKDE **24**(3), 533–546 (2012)

56. Ranbaduge, T., Christen, P., Vatsalan, D.: Tree based scalable indexing for multi-party privacy-preserving record linkage. In: AusDM (2014)
57. Ranbaduge, T., Vatsalan, D., Christen, P.: Clustering-based scalable indexing for multi-party privacy-preserving record linkage. In: Cao, T., Lim, E.-P., Zhou, Z.-H., Ho, T.-B., Cheung, D., Motoda, H. (eds.) PAKDD 2015. LNCS (LNAI), vol. 9078, pp. 549–561. Springer, Cham (2015). https://doi.org/10.1007/978-3-319-18032-8_43
58. Ranbaduge, T., Vatsalan, D., Christen, P., Verykios, V.: Hashing-based distributed multi-party blocking for privacy-preserving record linkage. In: Bailey, J., Khan, L., Washio, T., Dobbie, G., Huang, J.Z., Wang, R. (eds.) PAKDD 2016. LNCS (LNAI), vol. 9652, pp. 415–427. Springer, Cham (2016). https://doi.org/10.1007/978-3-319-31750-2_33
59. Ranbaduge, T., Vatsalan, D., Christen, P.: Scalable block scheduling for efficient multi-database record linkage. In: ICDM. Barcelona (2016)
60. Randall, S.M., Ferrante, A.M., Boyd, J.H., Semmens, J.B.: Privacy-preserving record linkage on large real world datasets. JBI **50**, 205–212 (2014)
61. Schneider, D.A., DeWitt, D.J.: Tradeoffs in processing complex join queries via hashing in multiprocessor database machines. In: VLDB, pp. 469–480 (1990)
62. Schnell, R.: An efficient privacy-preserving record linkage technique for administrative data and censuses. Stat. J. IAOS **30**(3), 263–270 (2014)
63. Sehili, Z., Kolb, L., Borgs, C., Schnell, R., Rahm, E.: Privacy preserving record linkage with PPJoin. In: BTW Conference, Hamburg (2015)
64. Shannon, C., Weaver, W.: The Mathematical Theory of Communication, vol. 19. University of Illinois Press, Urbana (1962)
65. Sweeney, L.: Computational disclosure control: A Primer on Data Privacy Protection. Ph.D. thesis, Massachusetts Institute of Technology, Department of Electrical Engineering and Computer Science (2001)
66. Tran, K.N., Vatsalan, D., Christen, P.: GeCo: an online personal data generator and corruptor. In: ACM CIKM, San Francisco, pp. 2473–2476 (2013)
67. Vatsalan, D., Christen, P.: An iterative two-party protocol for scalable privacy-preserving record linkage. In: AusDM, CRPIT, vol. 134, Sydney (2012)
68. Vatsalan, D., Christen, P., Verykios, V.S.: An efficient two-party protocol for approximate matching in private record linkage. In: AusDM, Ballarat (2011)
69. Vatsalan, D., Christen, P., Verykios, V.S.: A taxonomy of privacy-preserving record linkage techniques. JIS **38**(6), 946–969 (2013)
70. Vatsalan, D.: Scalable and approximate privacy-preserving record linkage. Ph.D. thesis, Research School of Computer Science, The Australian National University (2014)
71. Vatsalan, D., Christen, P.: Sorted nearest neighborhood clustering for efficient private blocking. In: Pei, J., Tseng, V.S., Cao, L., Motoda, H., Xu, G. (eds.) PAKDD 2013. LNCS (LNAI), vol. 7819, pp. 341–352. Springer, Heidelberg (2013). https://doi.org/10.1007/978-3-642-37456-2_29
72. Vatsalan, D., Christen, P.: Scalable privacy-preserving record linkage for multiple databases. In: ACM CIKM, Shanghai (2014)
73. Vatsalan, D., Christen, P., O'Keefe, C.M., Verykios, V.S.: An evaluation framework for privacy-preserving record linkage. JPC (2014)
74. Vatsalan, D., Christen, P., Rahm, E.: Scalable privacy-preserving linking of multiple databases using counting bloom filters. In: IEEE ICDMW, Barcelona, Spain (2016)

75. Vatsalan, D., Christen, P., Verykios, V.S.: Efficient two-party private blocking based on sorted nearest neighborhood clustering. In: ACM CIKM, San Francisco, pp. 1949–1958 (2013)
76. Vatsalan, D., Sehili, Z., Christen, P., Rahm, E.: Privacy-preserving record linkage for big data: current approaches and research challenges. In: Zomaya, A.Y., Sakr, S. (eds.) Handbook of Big Data Technologies, pp. 851–895. Springer, Cham (2017). https://doi.org/10.1007/978-3-319-49340-4_25
77. Vernica, R., Carey, M.J., Li, C.: Efficient parallel set-similarity joins using MapReduce. In: Proceedings of ACM SIGMOD, pp. 495–506 (2010)
78. Wang, G., Chen, H., Atabakhsh, H.: Automatically detecting deceptive criminal identities. Commun. ACM **47**(3), 70–76 (2004)
79. Whang, S.E., Menestrina, D., Koutrika, G., Theobald, M., Garcia-Molina, H.: Entity resolution with iterative blocking. In: ACM SIGMOD, Providence, Rhode Island, pp. 219–232 (2009)
80. Zezula, P., Amato, G., Dohnal, V., Batko, M.: Similarity Search: The Metric Space Approach, vol. 32. Springer, New York (2006). https://doi.org/10.1007/0-387-29151-2

# Web Frameworks Metrics
# and Benchmarks for Data Handling
# and Visualization

Alexandros Gazis$^{(\boxtimes)}$ and Eleftheria Katsiri

Department of Electrical and Computer Engineering, 67100 Xanthi, Greece
{agazis,ekatsiri}@ee.duth.gr

**Abstract.** This paper presents benchmarks regarding a web application that aims at displaying and visualizing a dataset for air quality monitoring, experimenting using two different programming languages. Specifically, an application is developed via the use of PHP and Python frameworks in order to study the impact of the CPU, the hard disk architecture and the operating system between each system. Detailed tests have been conducted regarding the necessary computing resources as well as the use of the network (delay, CPU usage etc.) for different operating systems and hardware specifications.

**Keywords:** Cloud computing · Wireless network · Django · Flask ·
PHP-FPM · Lamp · Unix · Linux · Web frameworks · Middleware ·
Network performance modeling · Network simulations ·
Data visualization time · CPU time · User time · Wall time ·
Max resident · Server hardware · Benchmarks

## 1 Introduction

In the era of the Internet of Things, a wide range of programming language frameworks have surfaced for developing applications in several platforms. While there is a wide range of different languages and tools based on the application's specific and the offered services, the main programming language for most websites is PHP (back-end). PHP offers a well-tested and documented path to create a web site from scratch right of the box. Although several new languages have been released over the last decade (e.g. Ruby), PHP use steadily remains in top ranks, powering more than 50% of the Web. According to the TIOBE index [1], there is a new trend regarding the use of Python for various applications, one of which is web development [2].

In order to assess the status quo, a Web application was developed with a similar presentation layer (front-end interface), using different available technologies. This application may display a dump of a database and a visualization of a day's measurements. The developed site was kept to bare minimum Html, CSS and JavaScript (mainly for producing the necessary graphs) as the

© Springer Nature Switzerland AG 2019
Y. Disser and V. S. Verykios (Eds.): ALGOCLOUD 2018, LNCS 11409, pp. 137–151, 2019.
https://doi.org/10.1007/978-3-030-19759-9_9

tests emphasized on the features of the server and thus future cloud service. Although the developed application used data regarding Air Quality Monitoring, the results presented in this paper can be expanded to various small or medium sized web applications, such as lists of appointments, logs, catalogs and to-do lists. Other real-life examples include online market places (e.g. a food delivery site consisting of order_id, quantity, food_list, after_tax_price), office appointments (timestamp, customer_name, further_description), parking ticket systems (timestamp, client_id, ticket_duration) and local libraries (isbn, title, authors, user_comments). All these examples are consisted of a relation (table) of several tuples (rows), which have less than five attributes (columns) per tuple.

Our objective was to measure the time response while loading and visualizing data that is big in volume and sparse in Tuples. The used dataset includes several measurements (temperature, humidity) and was intentionally stored as an SQL database and as a Comma Separated Values (CSV) file. The two formats aimed at a better understanding and a detailed comparison of the interactions of each developed method. In recent computer systems, the goal is set forward by creating native applications which are depended on the cloud, so as to offer the desired level of user experience. As a result, the main factors that impact the quality and robustness in this assessment were mainly associated with the system's performance and response.

## 2   Literary Review

The recent developments regarding the use of Internet of things as well as the introduction of new methods and techniques have resulted in various Web frameworks and implementations in order to produce a Web application [3]. Moreover, as mentioned in [4,5] the new era of Industry 4.0 generates a high demand to test and categorize the best practices and methods regarding their cost and user experience [6]. Nowadays, the design and implementation of a web application is more complex, due the size of the available data ("Big Data") [7,8]. In most cases, the developed web applications may not handle the provided information. Taking this into account we studied several papers presenting ways of measuring data handling, data visualization and time measurements with regard to the operating system and CPU utilization for the server [9]. In order to give an overview of the web programming languages, already utilized frameworks and architectures [10] were taken into consideration as well as a published paper regarding Network and Server Performance Evaluation [11]. Lastly, we conducted an extensive study to develop an application regarding Python's and PHP's implementations, which greatly impacted the attributes we analyzed (see, such as: [12]).

# 3  Aims and Objectives

The proposed method regarding the conducted experiments presents the results in a two folded stage:

1. Utilization and Duration Time in terms of CPU - Hard Disk Architecture.
2. Hardware benchmarks and Duration Time in terms of Operating system - available System Resources.

The first stage of experiments, except for CPU utilization, emphasizes the following time attributes:

1. User time: how long the CPU is outside the kernel but within a process.
2. System time: how long the CPU is inside the kernel, but within a process.
3. CPU time (total): sum of User and System time, counting the total duration of time for the CPU process.
4. Wall time: duration of real time, from the start of the call till the end (a program's execution within the CPU -high utilization- results for Wall and CPU time to have the same values).
5. Max resident: real size (not swapped) of available RAM resources, strictly committed to the process.

The second stage of experiments focuses on the following time attributes, regarding the launch of the web application from the browser:

1. Recovery & presentation time, of all data in text format.
2. Recovery & Visualization time, required to create a graphic of the selected records (e.g. a day's measurements).

These are based on the comparison between different disk architectures (solid state and hard disk drive) for processing a data set of 42.702 records, from a relational database management system (MySQL [13]) and a CSV file. These files are both stored locally on the "/var" directory for Linux experiments, as well as on "C:/test" directory, regarding Windows Operating System experiments. A flow chart representing the sequence of events for the 2 stages of experiments is shown in Fig. 2.

More specifically, the properties of the examined system are the following: Kernel 4.13.0-38, architecture x86-64 bit, Ubuntu 16.04 LTS, Intel(R) Core(TM) i7-7700HQ CPU @ 2.80 GHz, RAM 8 GB, HDD ST1000LM035-1RK1 1 TB 5400 rpm, SSD Samsung Electronics Disk 256 GB. Moreover, PHP is executed with the following specifications: nginx version: nginx/1.10.3 (Ubuntu), PHP 7.0.28-0ubuntu0.16.04.1 (fpm-fcgi) [14]. Respectively, Python's main packages used are the following: configparser (3.5.0), Django (2.0.4) [15], Flask (0.12.2) [16], httplib2 (0.9.1), itsdangerous (0.24), Jinja2 (2.10), louis (2.6.4), lxml (3.5.0), Mako (1.0.3), MarkupSafe (0.23), mysqlclient (1.3.12), Pillow (3.1.2), pip (9.0.3), setuptools (39.0.1), virtualenv (15.0.1), Werkzeug (0.14.1).

# 4    First Stage of Tests

## 4.1    Proposed Method

The first stage of tests was set in Linux Mint and aimed at comparing different hard disk architectures, while interacting with a web site. The experiment involved developing a simple website, which displayed all measurements, for a time span of certain weeks. In particular, three types of data were available; timestamp, temperature and humidity attributes. This site was executed locally, in our computer network, on a local host and summed up 6 different scenarios of web framework implementations, which collected the records from MySQL or a CSV file.

As for the architecture of the web app, our main focus was to test the most widely used frameworks in the field. Generally, there is a vast variety concerning Python Frameworks and the most used ones is Django and Flask. The first is undoubtedly one of the highest-level frameworks as it uses the Model-View-Template architecture. This framework enables rapid development of secure and maintainable websites due to its abstraction layer, availability of third-party libraries and proposed design pattern (Model-View-Control). In contrast, the second is mainly used as a tool for fast prototyping and doesn't offer pre-built functions and methods out of the box but requires from the user to import them. In other words, it provides a simple but extensible solution which doesn't recommend or force the user to develop a web application in a certain pattern of design (e.g. usually all the web server is stored in one .py file). Finally, as for PHP's framework we selected LAMP stack which is a group of open source software tools. More specifically, the acronym stands for Linux (operating system), Apache (web server), MySQL (database) and PHP (procession of dynamic content) and when these technologies are combined they enable the user to develop a server to host web sites and applications.

The cases tested were the following: Python Django fetch from MySQL, Python Django fetch from CSV file, Python Flask fetch from MySQL, Python Flask fetch from CSV file, PHP[php-fpm] fetch from MySQL, PHP fetch from CSV file. When database's records are displayed, a web page, approximately 2.6 (MB), is generated. In order to achieve these, the curl and time commands were executed in the Linux terminal (bash shell). The first set of commands served as a tool which incorporated the necessary -network- protocols to transfer data from or to a server. In our case study the command was used in order to send multiple HTML requests, thus acting as if users were browsing through our page. The second set was responsible for providing the following attributes: User time, System time, CPU time (total), Wall time and Max resident.

It is noted that due to the complexity of moving the location of the database for each test, in HDD tests, only the location of the CSV file and the executed scripts which provide the measurements should be altered. The produced results are generated from the execution of a script which initiates 10 requests each time (in a serial manner) and stores the values in a txt file. This number of requests was selected as the optimal value, after several tests to filter out any possible outliers. Although, this number should be as high as possible in theory (e.g. 1000) it is not significant in our case because any possible outlier will be detected during the first seconds of the test. Additionally, it is noted that a short value of time such as 1 (ms) or less won't affect the process of locating the outliers, due to the fact that more time is needed for the script to be executed inside the terminal. Each script is executed several times in order to minimize potential errors, while simultaneously executing the stress command and measuring the outcome of each iteration, via the top command. When several users are connected to the site the command siege [17] is used, which allows the simulation of a scenario, where multiple user requests are made enabling us to test the capabilities of the server (i.e. the number of handled requests per second).

Moreover, no tests were conducted regarding x86-32 bit architectures, due to recent developments in technology, since even embedded systems like Raspberry Pi [18] rely on x64-64 bit. Accordingly, the RAM's size does not affect the footprint of our cases because it is of small size and it does not affect the overall system. It should be noted that the available RAM resources are noticeable, only when a transfer occurs from RAM to swap. During our tests in that scenario, the computer system supporting the server decreases its response time to unacceptable levels.

Lastly, the operating system slightly affects the measurements, due to the use of CPU scaling. Several times, the CPU is not executed in maximum speed and the frequency of the cores is adjusted, depending on the load. Finally, it should be noted that this attribute is not connected with CPU time (total) measurement, in any way.

## 4.2   Model of Study

The values that are of outmost importance is Max resident and Wall time. These values can be generated to calculate the overall cost for developing, maintaining and scaling out our application in the server, thus improving the use of the host's cloud service. The first attribute is connected to the capabilities of our system to serve the user's request (without using swap), while the second is linked to the instruction cycles. In other words, these values are an indicator of the necessary capital and computer power to operate online.

The results regarding measurements of time, via executing the curl command in SSD-HDD, for 6 proposed cases, are presented in Tables 1, 2 and 3, 4 respectively.

Table 1 presents the time (calculated in seconds) regarding 10 iterations where an extremely high CPU usage occurs. From this table, we notice that methods using Python are optimal and less resource consuming than PHP. The explanation lies in the use of nginx as a web server by PHP [19]. As for the CPU util attribute which exceeds 100%, this value is not false, it is referring to an implementation where more than one core is needed for the execution (Hyper-threading technology).

Table 2 shows similar values with the previous table in the CPU util value. In this case, additional time is required in order to start and load the needed compilers, interpreters and libraries. The optimal choice in this scenario is Python implementations. Even though, PHP requires less memory in order to achieve the same results as Flask, the wall time difference is significant since it stops the execution of the web application, if it is in high values. As a result, the application needs to wait idly for the computer system to end up processing in order to continue its course of action.

Table 3 notes that there is not a substantial difference between the values of time even for the CSV file, which is unusual. This is because the Operating System's functionality and more specifically the caching phenomenon that occurred in Hardware level. As for the database there is no difference in the expected values due to the fact that the location where the records are stored has not altered during the test.

Table 4 indicates that, unexpectedly, Python's frameworks don't present an optimal solution for the proposed web application. As a result, these measurements are connected with the Input-Output of each operation in the compiler of PHP language and thus are extended to the execution of MySQL fetching.

Lastly, the first stage of tests was concluded with the use of a script that run the siege command for 10 s and 20 parallel connections (benchmark mode). In order to evaluate the extracted quantities, "news247" [20] i.e. one of the largest Greek news site, as well as "Twitter" [21] i.e an online news and social networking service, was used as a reference point. The results are presented in Table 5 and provide a unique insight, presenting a detailed overview of the system's properties. More specifically [17]:

1. Concurrency: average simultaneous connections (value increases as server performance decreases);
2. Data transferred: sum of transferred data for every simulated user (header information included);
3. Transactions: server hits (redirections & authentication challenges count as two hits);
4. Response time: average server time to respond to each simulated user's requests.

Specifically, the implementations of PHP programming language that make use of nginx cannot be compared with Python's, in case of multiple parallel simultaneous user requests. As evident from our results, PHP may handle 5 additional requests per second, in this case study. The explanation lies in the fact that Python is single-thread and even thought Django can be executed using multiple threads this implementation requires many configurations, which are not worth setting up for simple projects, such as the minimum size web page. As a result, this table shows the measurements for one connection per user of the page and thus loses its attribute of comparison due to lack of the parallel connections. The conclusion reached from the set of tests is that Python web frameworks are faster in cases of one connection. More importantly, the needed resources range in the same levels.

## 4.3  Results

To sum up, when fetching data from a CSV file - Database, a significant difference does not occur. It should be noted that the use of MySQL offers optimal results out of the box, when we alter data type accordingly. In addition, MySQL offers far more options to search records, whereas in a txt file, the same process is extremely slow and sluggish thus, the use of MySQL reduces the overall volume of data to be transmitted over the network.

Moreover, another important factor to take under consideration is the complexity of the installation. This fact extends to the language, as well as the needed tools in order to set up a server (locally) to host our website. In particular, Flask is easier to install, up to the point of running our full-stack application, especially in comparison to Django framework. The reason for this is that Flask simply encapsulates the whole application, visualization and site options in a single file (.py). In addition, Django framework offers far more features when it is set up but this is accurate only if we follow the conventions for development, as mentioned in the official documentation [22]. The recommended methods and patterns require a great deal of work and familiarity in order to rapidly create the necessary folders, sub-folder, etc. in compliance to each site's function. On the contrary, PHP is simple to install and implications only occur to the properties of nginx, where technical knowledge is necessary. Subsequently, it is proposed to use nginx, due to the fact that both Django and nginx are characterized by a better performance for static data (images, JavaScript files etc.).

Finally, a single programming language and by extension a framework cannot be proposed, in order to produce the optimal web application. The above-mentioned results show that each case scenario need to be thoroughly examined, in order to select the best tool. In general terms, when there is a need to develop a web application for a simple model of study (e.g. to present some measurements, sensor values etc.) Python Programming Language should be selected over PHP. Moreover, if front-end presentation of the site is not a priority, the use of Flask framework is proposed, as it is quick to set up and deployed. In addition, if there is a demand for developing a project on which several different versions and software engineers will collaborate, then Django would be the optimal choice. This lies in the fact that seemingly time spending procedures (i.e. creation and storage of files in specific locations, general rules for naming files and functions) are part of the overall philosophy around the Model-View-Template architecture. These patterns are used in order to develop, maintain and scale up an application. They also facilitate the team of engineers to understand and elaborate in a cost-efficient manner. As for the format of the file where our data is stored, if there is a need to perform a search or sorting of our records, then a database is by far considered a better solution than a CSV file.

## 5   Second Stage of Tests

### 5.1   Proposed Method

The second stage of tests was set in a virtual machine where several different operating systems and computer architectures were thoroughly examined. In this stage, the aim was to compare and assess the Duration of Time, in terms of Operating systems and available System Resources. The hardware benchmarks were provided via the use of the same data set, in CSV format and SQLite [23], instead of MySQL. The change in SQLite, an embedded type of database, from the "traditional" choice of the previous stage of test, was implemented in order to further shrink the overall system size. This database is used vastly in measurements regarding the Internet of Things, because it is: easily extended for all types of electronic devices and capable of rapidly importing and exporting raw data to and from CSV files. In addition, it provides fast and reliable data services to applications of low or medium traffic websites, which have fewer than 100 K hits per day [24]. In this stage, we created a virtual machine for each test with specific hardware specifications and we made a clean installation of each language and the packages/frameworks, then we executed a full database dump, as well as a day's measurement visualization. The aim was to develop a sandbox environment for each scenario and to extract measurements for each process, regarding the attribute of time. The measurements presented in the following sections were extracted and validated with two different tools, for each tested operated system (for Linux, via the use of system monitor & Glances [25] and for Windows via task manager & Process Explorer [26]).

## 5.2   Model of Study

This stage of experiments was consisted of 3 attributes from a database. The aim was to dump all the available records in a website, display a day's records (approximately 200 measurements) and visualize all the above. We selected the visualization of a day's records because there is no practical use of visualizing all the available data in a dataset of this size. This time, all tests were executed from a SSD, in order to obtain maximum read and write speeds for the performed tests. The visualization was consisted of a day's data as shown in Fig. 1.

## 5.3   Results

Firstly, using the process-time() function for Django Framework version 2.0.5 and Python 3.6.5, we acquire the necessary measurements from Google Chrome 66.0.3359.139 [27]. Secondly, through the same tests, we extracted results, regarding Flask Framework version 1.0.2 and Python 3.6.5. As for the PHP implementation, lamp software bundle was used which is a model of several web service stacks. The ones used were: PHP 7.2.5/7.1.17/7.0.30/5.6.36. In addition, through these tests, we did not note measurements for the boot time of each framework, because apache2 is executed as a service in the Operating System. Results from a day's measurements are presented in Table 6, as well as an indicative case study in Fig. 3, regarding the Retrieval and Display time for a mini PC of 2 Cores and 2/4 [GB] of RAM.

## 5.4   Results

A thorough analysis was made regarding a real, low budget system, which could be used locally as a server to host our web application. Several tests were made requiring that system which run Linux, which were consisted of different software architectures and computer systems. The above tables are important, so as to compare PHP and Python, as well as to assess and calculate the necessary budget for power and equipment, depending on our needs, in order to host an application online. As far as the boot time for Windows 10 is concerned, regardless of the architecture, Python Frameworks implementation is preferable, i.e. the ratio of speed performance is approximately 3 to 1. Moreover, the test case of multiple cores is interesting, because the above ratio for Python remains in a value range of similar levels, as for PHP. In addition, the PHP tests for Unix operating systems show that except for boot time, there is not a great dependency of the overall performance over the number of available CPU cores in the system.

As for the tests conducted regarding the visualization process for a day's records in Windows 10, while Python frameworks time measurements range between similar values, in PHP implementation, if more resources are provided -specifically increasing RAM size-, a decrease in the time is noted, so as to retrieve and visualize the available information. This, is very important, as it decreases the waiting time for each page request, thus leading to a smoother user experience. In the last tests, where we increased the provided RAM size, while the RAM's size increases, the PHP and Python frameworks tend to acquire the same system performance. Based on the above tables, for a system with 2 cores minimum and 4 GB of RAM, the framework does not affect the measurements regarding the time attributes, as there is no fluctuations between its values. Accordingly, Unix time measurements are generally similar to Windows, except for the Flask framework, where there is a small improvement in the user's response time (for a small amount of RAM, with max availability: 1 GB).

Moreover, regarding the CPU utilization, for various RAM sizes, similar results are drawn from all of our tests. Likewise, a change in RAM's size is not observed, if there is an alternation in the available CPU cores, which was an expected behavior. As for the tests of CPU utilization in Windows, there is a big difference between the 3 examined frameworks. Python is preferable regarding all CPU % consumption, regardless of the RAM's size availability, whereas PHP frameworks are less effective and far more power demanding. Thus, we can state that the use of a system like Raspberry Pi, which was the original system hosting our website is a relatively optimal choice, in order to achieve good results, as the 1 GB of RAM and the overall CPU % consumption is low, apart from PHP. The benchmarks, regarding the use of Raspberry Pi as a server, running a Debian-based operating system, Raspbian [29], are presented in Table 7. This is very important in order to select and calculate the future pricing of a cloud service, i.e. the costs of future scale up.

## 6   Future Work

A simple web application was developed and tested, in order to examine the impact of two programming languages and its mainstream used Frameworks. As for future recommendations, more tests for asynchronous requests could be added. Emphasis should be given on frameworks that strictly focus on the use of Model View Control (similar to Model-View-Template of Django) Architecture over PHP's Laravel [28] and Python's Django frameworks. By the same token, more micro frameworks, like Flask, can be put under the microscope, such as PHP's Silex [30] or Ruby's Sinatra [31] or Java's Spark [32]. In addition, it would be interesting to enhance our tests, via using more languages and frameworks, such as de facto enterprises' language Java and Spring [33], Ruby and by extension Ruby on Rails [34], as well as contemporary technologies, i.e. JavaScript Node.js [35] and Angular.js [36]. In addition, we could study multi-threading cases, which are necessary for implementations associated with data visualization and provide a tool or a web extension that could instantly execute the above-mentioned test stages [37].

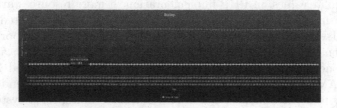

**Fig. 1.** Data visualization regarding a day's measurement of approximately 200 records

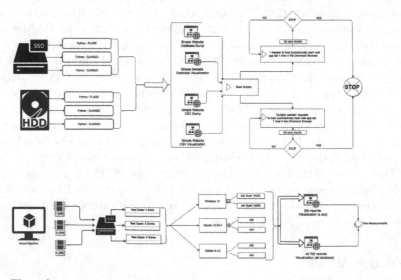

**Fig. 2.** Flow chart representing the sequence of events for the 2 stages of experiments

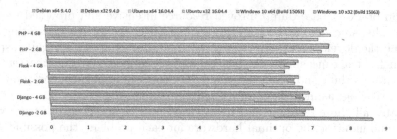

**Fig. 3.** PHP (LAMP) - Flask - Django Retrieval and Display measurements for the whole dataset regarding the case study of 2 cores and 2/4 GB of RAM (e.g. mini PC)

**Table 1.** SSD time duration measurements via executing curl command

|  | Django | | Flask | | PHP | |
|---|---|---|---|---|---|---|
|  | MySQL | CSV | MySQL | CSV | MySQL | CSV |
| User time | 0.02 | 0.04 | 0.02 | 0.03 | 0.04 | 0.03 |
| System time | 0.02 | 0.01 | 0.02 | 0.02 | 0.05 | 0.06 |
| CPU time (total) | 0.04 | 0.05 | 0.04 | 0.05 | 0.09 | 0.09 |
| Wall time | 0.05 | 0.05 | 0.04 | 0.05 | 0.7 | 0.67 |
| Max resident [MB] | 7.6 | 7.6 | 7.6 | 7.6 | 7.6 | 7.6 |
| CPU util | 98% | 101% | 102% | 91% | 13% | 15% |

**Table 2.** SSD time duration measurements via executing direct requests

|  | Flask | | PHP | |
|---|---|---|---|---|
|  | MySQL | CSV | MySQL | CSV |
| User time | 2.56 | 2.27 | 1,09 | 1.25 |
| System time | 0.24 | 0.28 | 4.9 | 4.66 |
| CPU time (total) | 2.8 | 2.55 | 5.99 | 5.91 |
| Wall time | 2.82 | 2.56 | 6.08 | 5.98 |
| Max resident [MB] | 54.6 | 54.6 | 47.1 | 47.0 |
| CPU util | 99% | 99% | 98% | 98% |

Moreover, during the first stage of experiments, regarding attributes of computer networks, we noticed strange results explained by the use of nginx over the PHP implementation. It would be interesting to re-run all the above and use nginx or extract measurements regarding WSGI and uWSGi with Nginx in Flask's and Django's framework. Similar tests can be expanded to bigger data sets that would use Apache Hadoop [38], Apache Spark [39], High-Performance Computing Cluster and Google BigQuery [40]. In addition, an extensive study can be conducted, regarding the enhancement of users' experience operating on small computer systems, such as Raspberry Pi. For example, simple tasks, as the provision of a database dump, proved resource-intensive and provided unsatisfying results, since the system had free (idle) CPU cycles. An interesting task would be the development of a networked computer grid, e.g. several Raspberries, which extract measurements (slaves) and are connected to one (master) that solely analyzes data and hosts the server [41], so as to implement techniques for CPU scavenging, as mentioned in: [42,43].

Finally, all these results could be modeled in a Neural Network that would predict and monitor the optimal hardware for each problem, the possible selections regarding data provided by the user, the cost to scale up an application in the cloud (mainly for the CPU % consumption) if it exceeds the capabilities of our server and more importantly the ratio, regarding the cost of hardware and high cloud service standards [44,45].

**Table 3.** HDD time duration measurements via executing curl command

|  | Django | | Flask | | PHP | |
| --- | --- | --- | --- | --- | --- | --- |
|  | MySQL | CSV | MySQL | CSV | MySQL | CSV |
| User time | 0.01 | 0.02 | 0.03 | 0.03 | 0.03 | 0.07 |
| System time | 0.03 | 0.02 | 0.01 | 0.02 | 0.08 | 0.04 |
| CPU time (total) | 0.04 | 0.04 | 0.04 | 0.05 | 0.11 | 0.11 |
| Wall time | 0.05 | 0.05 | 0.04 | 0.37 | 0.59 | 0.59 |
| Max resident [MB] | 7.6 | 7.6 | 7.6 | 7.6 | 7.6 | 7.6 |
| CPU util | 104% | 104% | 104% | 17% | 16% | 19% |

**Table 4.** HDD time duration measurements via executing direct requests

|  | Flask | | PHP | |
| --- | --- | --- | --- | --- |
|  | MySQL | CSV | MySQL | CSV |
| User time | 2.36 | 2.33 | 3.4 | 3.24 |
| System time | 0.24 | 0.28 | 27.84 | 27.5 |
| CPU time (total) | 2.6 | 2.61 | 31.23 | 30.74 |
| Wall time | 2.73 | 2.75 | 95.58 | 94.11 |
| Max resident [MB] | 54.6 | 54.6 | 47.1 | 47.0 |
| CPU util | 99% | 99% | 32% | 32% |

**Table 5.** Required resources by the computer system during siege execution

|  | Flask | | Django | | PHP - SSD | | PHP - HDD | | news247.gr | twitter.com |
| --- | --- | --- | --- | --- | --- | --- | --- | --- | --- | --- |
|  | MySQL | CSV | MySQL | CSV | MySQL | CSV | MySQL | CSV | | |
| Concurrency | 17.19 | 17.34 | 17.19 | 17.24 | 19.49 | 19.50 | 19.46 | 19.58 | 18.51 | 19.06 |
| Data transferred [MB] | 182.19 | 138.57 | 182.19 | 138.57 | 987.97 | 1098.32 | 952.04 | 1159.91 | 21.71 | 6.25 |
| Transactions [hits] | 73 | 53 | 71 | 54 | 385 | 428 | 371 | 452 | 184 | 240 |
| Response time [sec] | 2.25 | 3.19 | 2.25 | 3.19 | 0.51 | 0.46 | 0.52 | 0.43 | 1.01 | 0.79 |

**Table 6.** Django-Flask-PHP web stack(lamp) frameworks Retrieval and Visualization (regarding Fig. 1) time measurements for a day's records

| | Django 1 Core | | | Django 2 Cores | | | Django 4 Cores | | | Flask 1 Core | | | Flask 2 Cores | | | Flask 4 Cores | | | PHP 1 Core | | | PHP 2 Cores | | | PHP 4 Cores | | |
|---|---|---|---|---|---|---|---|---|---|---|---|---|---|---|---|---|---|---|---|---|---|---|---|---|---|---|---|
| *Test Results for 1,2,4 [GB] of RAM* | | | | | | | | | | | | | | | | | | | | | | | | | | | |
| Windows 10 x32 (Build 15063) | 3.44 | 2.99 | 2.94 | 3.20 | 3.01 | 2.90 | 2.43 | 2.23 | 2.17 | 3.28 | 2.86 | 2.82 | 3.03 | 3.00 | 2.73 | 2.43 | 2.23 | 2.14 | 3.89 | 2.99 | 2.93 | 3.48 | 3.47 | 3.44 | 2.84 | 2.80 | 2.76 |
| Windows 10 x64 (Build 15063) | - | 3.14 | 2.99 | - | 2.78 | 2.56 | - | 2.45 | 2.20 | - | 3.10 | 2.87 | - | 2.61 | 2.38 | - | 2.44 | 2.14 | - | 3.71 | 3.53 | - | 3.06 | 2.85 | - | 2.77 | 2.71 |
| Ubuntu x32 16.04.4 | 3.33 | 3.01 | 3.0 | 3.23 | 3.20 | 3.18 | 3.23 | 3.21 | 3.20 | 3.22 | 3.17 | 3.15 | 3.15 | 3.06 | 3.02 | 3.29 | 2.89 | 2.84 | 3.83 | 3.60 | 3.30 | 3.73 | 3.53 | 3.26 | 3.32 | 3.26 | 2.94 |
| Ubuntu x64 16.04.4 | - | 3.04 | 3.0 | - | 2.45 | 2.43 | - | 2.39 | 2.14 | - | 3.00 | 2.88 | - | 2.88 | 2.25 | - | 2.38 | 2.08 | - | 3.61 | 3.01 | - | 2.73 | 2.72 | - | 2.71 | 2.65 |
| Debian x32 9.4.0 | 3.27 | 3.20 | 3.20 | 3.20 | 3.19 | 3.19 | 3.19 | 3.19 | 3.17 | 3.18 | 3.13 | 3.14 | 3.23 | 3.08 | 3.04 | 3.03 | 3.16 | 3.01 | 3.79 | 3.26 | 3.25 | 3.73 | 3.71 | 3.36 | 3.79 | 3.26 | 3.25 |
| Debian x64 9.4.0 | - | 2.99 | 2.93 | - | 2.44 | 2.37 | - | 2.30 | 2.24 | - | 2.95 | 2.81 | - | 2.27 | 2.19 | - | 2.29 | 2.18 | - | 3.56 | 2.94 | - | 2.66 | 2.72 | - | 2.75 | 2.63 |

**Table 7.** Raspberry Pi's benchmarks, regarding daily and overall measurements

| | Django | Flask | PHP |
|---|---|---|---|
| Boot time [sec] | 4.93 | 4.53 | 5.26 |
| Ram required (without background processes) [MB] | 27.9 | 22.1 | 49.5 |
| CPU usage at server launch | 27% | 19% | 32% |
| Retrieval time for 200 measurements (a day) [sec] | 1.08 | 0.89 | 1.13 |
| Retrieval time for the whole dataset [sec] | 25.21 | 24 | 27 |
| Visualization time for 200 measurements (a day) [sec] | 4.17 | 3.77 | 4.77 |
| Visualization time for the whole dataset [sec] | 57.08 | 56.05 | 59 |
| Maximum RAM required for 200 measurements (a day) [MB] | 37 | 29 | 56 |
| Maximum RAM required for the whole dataset [MB] | 53.2 | 53 | 61 |
| Maximum CPU usage for 200 measurements (a day) | 7.02% | 4% | 11.83% |
| Maximum CPU usage for the whole dataset | 24% | 27% | 34% |

# References

1. Tiobe index (2018). https://www.tiobe.com/tiobe-index/
2. Titchkosky, L., Arlitt, M., Williamson, C.: A performance comparison of dynamic web technologies. SIGMETRICS Perform. Eval. Rev. **31**, 2–11 (2003)
3. Shenker, S.: Fundamental design issues for the future internet. IEEE J. Sel. Areas Commun. **13**, 1176–1188 (1995)
4. Drath, R., Horch, A.: Industrie 4.0: hit or hype? [industry forum]. IEEE Ind. Electron. Mag. **8**, 56–58 (2014)
5. Hermann, M., Pentek, T., Otto, B.: Design principles for industrie 4.0 scenarios. In: 49th Hawaii International Conference on System Sciences (HICSS), pp. 3928–3937, January 2016
6. Wan, J., Tang, S., Shu, Z., Li, D., Wang, S., Imran, M., Vasilakos, A.V.: Software-defined industrial internet of things in the context of industry 4.0. IEEE Sens. J. **16**, 7373–7380 (2016)
7. Vigo, M., Brajnik, G.: Automatic web accessibility metrics: where we are and where we can go. Interact. Comput. **23**, 137–155 (2011)

8. Matias, Y.: On big data algorithmics. In: Epstein, L., Ferragina, P. (eds.) ESA 2012. LNCS, vol. 7501, p. 1. Springer, Heidelberg (2012). https://doi.org/10.1007/978-3-642-33090-2_1

9. Dhyani, D., Ng, W.K., Bhowmick, S.S.: A survey of web metrics. ACM Comput. Surv. **34**(4), 469–503 (2002)

10. Jailia, M., Kumar, A., Agarwal, M., Sinha, I.: Behavior of MVC (model view controller) based web application developed in PHP and .NET framework. In: International Conference on ICT in Business Industry Government (ICTBIG), pp. 1–5, November 2016

11. Barford, P., Crovella, M.: Generating representative web workloads for network and server performance evaluation (1997)

12. Walker, D., Orooji, A.: Metrics for web programming frameworks. In: International Conference on Semantic Web and Web Services (2011)

13. Mysql (2018). https://www.mysql.com/

14. Php-fpm (2018). https://www.php-fpm.org

15. The web framework for perfectionists with deadlines (2018). https://www.djangoproject.com

16. Welcome—flask (a python microframework) (2018). http://www.flask.pocoo.org

17. Siege(1): HTTP/HTTPS stress tester - Linux man page (2018). https://linux.die.net/man/1/siege

18. Raspberry pi-teach, learn and make with raspberry pi (2018). https://www.raspberrypi.org

19. Nginx—high performance load balancer, web server, reverse (2018). https://www.nginx.com

20. News247 (2018). https://www.news247.gr

21. Twitter. It's what's happening (2018). https://twitter.com

22. Django documentation (2018). https://docs.djangoproject.com/en/2.0/

23. Sqlite home page (2018). https://www.sqlite.org/

24. Appropriate uses for sqlite (2018). https://www.sqlite.org/whentouse.html

25. Glances - an eye on your system (2018). https://nicolargo.github.io/glances/

26. Process explorer — windows sysinternals — microsoft docs (2018). https://docs.microsoft.com/en372us/sysinternals/downloads/process-explorer

27. Chrome web browser - Google (2018). https://www.google.com/chrome/

28. Laravel-the php framework for web artisans (2018). https://www.laravel.com

29. Raspbian (2018). https://www.raspbian.org

30. Silex - the php micro-framework based on the symfony (2018). https://silex.symfony.com

31. Sinatra: Classy web-development dressed in a dsl (2018). https://github.com/sinatra/sinatra

32. Spark Framework: An expressive web framework for Kotlin and Java (2018). http://sparkjava.com

33. Spring (2018). https://spring.io

34. Ruby on rails — a web-application framework that includes everything (2018). https://rubyonrails.org

35. Http — node.js v10.6.0 documentation (2018). https://nodejs.org/api/http.html

36. Angular (2018). https://angular.io

37. Gutwin, C.A., Lippold, M., Graham, T.C.N.: Real-time groupware in the browser: testing the performance of web-based networking. In: Proceedings of the ACM Conference on Computer Supported Cooperative Work - CSCW, pp. 167–176. ACM, New York (2011)

38. Apache Hadoop (2018). https://hadoop.apache.org/
39. Apache Spark - Unified Analytics Engine for Big Data (2018). https://spark.apache.org/
40. BigQuery - Analytics Data Warehouse — BigQuery — Google Cloud (2018). https://cloud.google.com/bigquery/
41. Gazis, A., Stamatis, K., Katsiri, E.: A method for counting, tracking and monitoring of visitors with RFID sensors model of study: M. Hatzidakis residence. In: Proceedings of 10th Panhellenic Electrical and Computer Engineering Students Conference (ECESCON), pp. 199–204, March 2018
42. Calvo, I., Gil-García, J.M., Recio, I., López, A., Quesada, J.: Building IoT applications with raspberry Pi and low power IQRF communication modules. Electron. Raspberry Pi Technol. **5**, 54 (2016)
43. Martinez, B., Vilajosana, X., Chraim, F., Vilajosana, I., Pister, K.S.J.: When scavengers meet industrial wireless. IEEE Trans. Ind. Electron. **62**(5), 2994–3003 (2015)
44. Pal, S.K., Talwar, V., Mitra, P.: Web mining in soft computing framework: relevance, state of the art and future directions. IEEE Trans. Neural Netw. **13**, 1163–1177 (2002)
45. Elhadik, S., Desoky, A.: Cognitive performance application. In: 2017 IEEE International Symposium on Signal Processing and Information Technology (ISSPIT), pp. 317–324, December 2017

# Algorithms for Cloud-Based Smart Mobility

Kalliopi Giannakopoulou[1,2]([✉])

[1] Department of Computer Engineering and Informatics, University of Patras,
26504 Patras, Greece
[2] Computer Technology Institute and Press "Diophantus", 26504 Patras, Greece
gianakok@ceid.upatras.gr

**Abstract.** Innovative algorithm technology plays an important role in smart city applications. In this work, we review some recent innovative algorithmic approaches that contributed decisively in the development of efficient and effective cloud-based systems for smart mobility in urban environments.

## 1 Introduction

Developing personalized mobility services in urban environments that combine public transportation modalities (metro, bus, tram, light rail, etc.) along with the emergence of modern transportation models like bicycles, electro-mobility, and car sharing are considered vital in creating efficient renewable mobility chains for travelers. Such a development is characterized by a host of grand challenges that concern route planning in urban traffic networks as well as provision of innovative features of urban mobility.

Modern urban mobility requirements call for the development of integrated applications that will offer an urban traveler several innovative features. Some of the most prominent features include:

- *Renewable mobility on demand*: an end-user with a variety of requests, requiring a multitude of eco-friendly services, should be offered a host of intermodal mobility opportunities, including electric vehicle (EV) transportation options. EVs supplement public transportation systems, providing mobility for the first and last part of a trip, which means that an ideal deployment would be in urban car sharing fleets located near public transport hubs.
- *Integrated personal mobility*: support and provide intermodal mobility opportunities for individuals (with emphasis on environmentally friendly means of transportation, such as bicycles and electric transportation media), which give the traveler the possibility to combine various means of transportation in an efficient mobility-chain. This is in line with the next generation bike-sharing systems that are built in major European cities.
- *Information Updating*: the characteristics of real-world transport networks, apart from demonstrating a predetermined behavior, also have to cope with

Y. Disser and V. S. Verykios (Eds.): ALGOCLOUD 2018, LNCS 11409, pp. 152–168, 2019.
https://doi.org/10.1007/978-3-030-19759-9_10

unpredicted incidents (e.g., temporal blockages of road segments due to construction works, accidents, delays of trains or buses, etc.), which are typically reported by several sources of information (e.g., municipality, police, travelers themselves). This live-traffic information has to efficiently update the historic traffic information, so that the routing service provides actual routes to the travelers (updating, if necessary, their initial plan).

The efficient and effective support of these features entails the development of sophisticated route planning applications, whose core components have to lie on a cloud architecture, in order to guarantee data persistence, interoperability with other traffic-related information sources, and transparent accessibility by the travelers. In such applications, the route planning queries are sent, from the mobile device of a traveler, to a routing engine residing at some cloud infrastructure, which in turn sends back the answers to the mobile device. Apart from communication latency, service efficiency heavily depends on the response time of the routing engine. This is a highly non-trivial challenge when route planning concerns large-scale transportation networks. Consequently, the efficient development of cloud-based routing services depends on the one hand on the overall architectural setting, and on the other hand on the algorithmic efficiency of the routing engine.

In this work, we survey recent innovative algorithmic technology embedded in core routing engines of cloud-based mobility applications in urban environments that played a critical role in the efficient development of such services and their effective applicability in practice. A prime example concerns the applications developed in the frame of [11,20]. In particular, we review algorithms for routing in road networks and for computing itineraries in multimodal public transit networks.

## 2    Algorithms for Smart Mobility in Road Networks

An important characteristic of road networks is that their travel time metric is time-dependent. Hence, any routing engine that wishes to provide accurate arrival travel time estimations at a desired destination, when a traveler starts at a certain origin at a particular departing time, should take time-dependent travel times into account.

All algorithmic approaches for time-dependent route planning in road networks first pre-process the input network so that subsequently queries for any origin-destination pair is answered instantaneously. In this setting, two main approaches were appeared: (i) heuristic techniques that speedup query time responses, while providing empirical evidence of the quality of the returned solution; and (ii) succinct data structures (oracles) that answer queries fast, while delivering a provable guarantee on the quality of the returned solution.

Along this framework, we shall review in this section two leading algorithmic techniques. In particular, the technique in [2] that falls within the heuristic approach, and the techniques in [12,17–19] that fall within the oracle approach. We shall elaborate more on the latter due to its special feature of providing a quality guarantee in the returned solution. All algorithmic techniques constituted the core routing engines of many modules in an integrated cloud-based mobility application [11,20].

## 2.1  Time-Dependent Heuristics

The work in [2] is an empirical approach (speedup technique) based on an extension of the Customizable Route Planning (CRP) technique [6]. CRP offloads most preprocessing effort to a metric-independent, separator-based phase. Preprocessed data is then customized to a given routing (time-independent) metric for the whole network within seconds or below. This also enables robust integration of user preferences.

The approach in [2], called time-dependent CRP (TDCRP), carefully extends CRP to time-dependent functions. As such, TDCRP evaluates partition-based overlays on a challenging non-scalar metric by integrating profile search into CRP's customization phase and by computing time-dependent overlays. Unlike CRP, a naive extension fails, since shortcuts on higher-level overlays are too expensive to be kept in memory (and too expensive to evaluate during queries). To reduce functional complexity, TDCRP approximates overlay arcs. In fact, approximation subject to a very small error suffices to make the approach practical. The resulting algorithmic framework enables interactive queries with low average and maximum error in a very realistic scenario consisting of live traffic, short-term traffic predictions, and historic traffic patterns. Moreover, it supports user preferences such as lower maximum driving speeds or the avoidance of highways. In an extensive experimental setup, it is demonstrated that TDCRP enables integration of custom updates much faster than previous approaches, while allowing fast queries that enable interactive applications. It is also robust to changes in the metric that turn out to be much harder for previous techniques.

## 2.2  Time-Dependent Oracles

The work in [18,19] are the first approaches that provide provable approximation guarantees to time-dependent travel time estimations, as they is based on new theoretical methods. In particular, in [18,19] new structures (oracles) are provided which allow efficient preprocessing of the input network (provably subquadratic on its size) so that subsequently queries can be answered efficiently estimating the arrival time at destination with a provable approximation guarantee.

The main idea is to select a proper subset of network nodes (*landmarks*) from which travel time summaries to all destinations are pre-computed. Then, the query algorithms exploit these summaries to provide approximate arrival travel-time estimates of a provably very small error.

The cloud-based structure of the time-dependent routing (TDR) oracles in [18,19] is described in [12]. The particular TDR engine provides fast and accurate shortest-path responses, exploiting a set of traffic-related data of three main types:

- Periodically updated raw traffic data (arc travel-time functions).
- Traffic metadata periodically produced in a preprocessing phase (landmark travel-time summaries provided by the time-dependent oracle).
- Live-traffic reports, that is, emergency alerts produced in an ad-hoc fashion by crowdsourcing and traffic-prediction mechanisms. Additionally, periodically updated snapshots of the traffic-related data are disseminated to the travelers' smartphone devices, in support of elementary routing services (as a contingency plan in case of connectivity loss with the cloud).

The TDR service supports three approximate query algorithms, which provide significantly fast and accurate min-cost route plan responses to arbitrary shortest-path queries, by exploiting the historical and temporal traffic-related information kept in the cloud-server. The TDR-daemon continuously runs and accepts incoming origin/destination/departure-time shortest-path queries $(o, d, t_o)$. For each one of them, a selected query algorithm is called that returns either the exact travel-time value corresponding to the exact $o$-$d$ path, or an approximate travel-time value via an appropriate landmark-node $\ell$, which corresponds to an approximate $o$-$\ell$-$d$ path. The query algorithms are as follows.

- *FCA*: it grows a Time-Dependent-Dijkstra (TDD) ball[1] from $(o, t_o)$ until either $d$ or the closest landmark $\ell_o$ is settled. It then returns either the exact solution, or an $(1 + \varepsilon + \psi)$-approximate value via $\ell_o$, where $\varepsilon$ is a small constant imposed by the preprocessing phase, and $\psi$ is another small constant depending on the characteristics of the travel-time functions but not on the size of the network.
- *FCA*$^+$(N): it is the default query algorithm provided by TDR. It is a variant of FCA which keeps growing a TDD ball from $(o, t_o)$ until either $d$ or a given number $N$ of landmarks is settled. FCA$^+$ then returns the exact travel-time value or the smallest via-landmark approximate travel-time value, among all $N$ settled landmarks. Theoretically, the approximation guarantee is the same as that of FCA, but the practical performance of FCA$^+$ is impressive.
- *RQA*: it exploits a number $r$ (called the recursion budget) of recursive accesses to the landmark-based preprocessed information, each of which produces (via calls to FCA) additional candidate $o$-$d$ paths. In that way, RQA improves the approximation guarantee provided by FCA. The idea of RQA is that as long as $d$ has not yet been discovered within the explored area around the origin, and there is still some remaining recursion budget, it "guesses" (by exhaustively searching for it) the next node of the boundary set of touched nodes (i.e., still in the priority queue) along the unknown $o$-$d$ path. At the end, it returns the best among all paths found.

An extensive experimental study of the TDR oracles in [18,19] has been carried out in [16], demonstrating their practicality on real-world road networks. The experimental study has also shown that in the vast majority of the queries the exact solution is discovered by all three algorithms.

---

[1] It runs TDD [10], a straightforward time-dependent variant of Dijkstra's algorithm, whose growing of search space (set of settled nodes) resembles the growing of a ball.

In the above oracle framework, a new time-dependent distance oracle (CFLAT) was presented in [17], which is the first oracle that preprocesses *combinatorial structures* (collections of time-stamped min-travel-time-path trees) rather than travel-time functions, and which exhibits an impressive practical performance.

CFLAT works as follows. In a preprocessing phase, it constructs and compactly stores min-cost-path trees at carefully sampled departure-times, rooted at each landmark $\ell$. This is achieved through a new approximation method, called CTRAP, whose novelty relies in exploiting the fact that there are significantly fewer changes in the combinatorial structure, than in the functional description of the optimal solution. Moreover, multiple copies of the same preprocessed information are avoided, by organizing the destinations from a landmark into groups of (roughly) equidistant vertices, for which the common departure-times sequence is stored only once.

A query $(o, d, t_o)$ is answered by algorithm CFCA($N$), which starts by growing a TDD ball from $o$ at time $t_o$, until either $d$ or the $N$ closest landmarks are settled. In the latter case, starting from $d$, a suitably small subgraph is constructed (consisting of certain paths going from $d$ back to $o$, using the settled landmarks as "attractors"), until a settled vertex of the initial TDD ball is reached. Then, a continuation of growing the initial TDD ball on the resulted small subgraph returns an $od$ path that turns out to approximate very well the optimal $od$ path. Note that CFCA computes the actual connecting path that preserves the theoretical approximation guarantees. To make it practical and tackle the main burden of landmark-based oracles (the large preprocessing requirements), CFLAT is extensively engineered.

The practical performance of CFLAT was assessed on two real-world benchmark instances, the urban area of Berlin and the national road network of Germany. In particular, for Berlin, using 16K landmarks ($N = 1$), a query time of 0.064 msec with a relative error of 0.0021 is achieved. For Germany, using 4K landmarks ($N = 1$), a query time of 0.057 msec with a relative error of 0.0078 is achieved. This query-time performance is competitive to state-of-the-art speedup heuristics for time-dependent road networks, whose query-times in most cases do not account for path construction.

# 3    Algorithms for Smart Mobility in Public Transit Networks

In this section, we review the algorithmic technology required to tackle efficiently the multimodal routing or journey planning problem in public transport networks. We consider multimodal schedule-based public transport (e.g., train, bus, tram) along with unrestricted w.r.t. departing time traveling (walking and electric vehicles – EVs).

The two most common optimization problems in multimodal journey planning are: (i) the *earliest arrival-time problem* (EAP), in which one is interested in finding the best (or optimal) journey that minimizes the traveling time required to complete it; (ii) the *minimum number of transfers problem* (MNTP), in which one is interested in computing a best journey that minimizes the number of times a passenger needs to change vehicle during the journey; (iii) the multicriteria version of these two (EA and MNT) optimization criteria.

For solving the aforementioned problems, we consider two prime approaches, both of which preprocess the input (multimodal) transport network so that subsequently queries are answered fast. The first one [5,8] is an array-based approach that puts more emphasis in providing fast answers to multimodal and multicriteria queries. The second one [3,4] is a graph-based approach that deals efficiently both wrt to fast query answering as well as with extremely fast updating of multimodal routing (itinerary) information in case of delays. For this reason (extremely efficient update routine in case of delays), we shall elaborate more on the second approach.

## 3.1 Array-Based Approaches

We start our exposition with the classical Multi-Label-Correcting (MLC) algorithm [22]. The MLC algorithm is a multicriteria shortest-path algorithm that finds full Pareto sets for arbitrary criteria that can be modeled as arc costs. MLC extends Dijkstra's algorithm by operating on labels that have multiple values, one per criterion. Each vertex $v$ maintains a bag $B(v)$ of non-dominated labels. In each iteration, MLC extracts from a priority queue the minimum (in lexicographic order) unprocessed label $L(u)$. For each arc $(u, v)$ out of the associated vertex $u$, MLC creates a new label $L(v)$ (by extending $L(u)$ in the natural way) and inserts it into $B(v)$. New dominated labels (possibly including $L(v)$ itself) are discarded, and the priority queue is updated if needed. MLC can be sped up with target pruning and by avoiding unnecessary domination checks.

The RAPTOR (Round bAsed Public Transit Optimized Router) algorithm [8] was introduced as a faster alternative to MLC for public transportation networks. The simplest version of the algorithm optimizes two criteria: arrival time and number of transfers. Unlike MLC, which searches a graph, RAPTOR uses dynamic programming to operate directly on the timetable. It works in rounds, with round $i$ processing all relevant journeys with exactly $i - 1$ transfers. It maintains one label per round $i$ and stop $p$ representing the best known arrival time at $p$ for up to $i$ trips. During round $i$, the algorithm processes each route once. It reads arrival times from round $i - 1$ to determine relevant trips (on the route) and updates the labels of round $i$ at every stop along the way. Once all routes are processed, the algorithm considers potential transfers to nearby (predefined) stops in a second phase. Simpler data structures and better locality make RAPTOR an order of magnitude faster than MLC. In [8] McRAPTOR was also proposed, which extends RAPTOR to handle more criteria (besides arrival times and number of transfers). It maintains a bag (set) of labels with each stop and round.

The aforementioned round-based paradigm (RAPTOR) has been adapted to the multimodal scenario in [5,9], where the MCR (multimodal multicriteria RAPTOR) was proposed that extends McRAPTOR to handle multimodal queries.

As in McRAPTOR, each round has two phases: the first processes trips in the public transportation network, while the second considers arbitrary paths in the unrestricted networks. They use a standard McRAPTOR round for the first phase (on the timetable network) and MLC for the second (on the walking network). Labels generated by one phase are naturally used as input to the other. During the second phase, MLC extends bags instead of individual labels. To ensure that each label is processed at most once, one keeps track of which labels (in a bag) have already been extended. The initialization routine (before the first round) runs Dijkstra's algorithm on the walking network from the source $s$ to determine the fastest walking path to each stop in the public transportation network (and to destination $t$), thus creating the initial labels used by MCR. During round $i$, the McRAPTOR subroutine reads labels from round $i - 1$ and writes to round $i$. In contrast, the MLC subroutine may read and write labels of the same round if walking is not regarded as a trip. In [5] a bike rental scheme is also considered, which can be seen as a hybrid network: it does not have a fixed schedule (and is thus unrestricted), but bikes can only be picked up and dropped off at designated bike stations. Picking a bike from its station counts as a trip. To handle bikes within MCR, they are considered during the first stage of each round (together with RAPTOR and before walking). Because bikes have no schedule, they are processed independently (from RAPTOR) by running MLC on the bike network. MLC is initialized with labels from round $i - 1$ for all relevant bike stations and, during the algorithm, labels of (the current) round $i$ are updated.

## 3.2    Graph-Based Approaches

The representation of graph-based information systems for scheduled-based transportation is described by timetables that determine the (scheduled) departure and arrival times of public vehicles.

A *timetable* is considered as a set tuple $T = (Z, B, C)$, where $B$ is the set of *stops* (or stations) in which the passengers may embark/disembark on/from a vehicle, $Z$ is the set of *vehicles* (train, bus, metro, and any other means of transport that performs scheduled routes), and $C$ is the set of *elementary connections* $c = (z, S_d, S_a, t_d, t_a)$, which represents the travel of a vehicle $z \in Z$, leaving from stop $S_d \in B$ at time $t_d$ and arriving at the immediately next stop $S_a \in B$ at time $t_a$. Elementary connections of scheduled-based transport are restricted w.r.t. (the scheduled) departure of the vehicles.

One of the most common models for representing a timetable is the *realistic time-expanded model* (TE-real) [24]. This model encodes a timetable $T$ into a directed graph $G = (V, E)$ with appropriate arc weights. In TE-real, nodes represent time events (arrival or departure times of a vehicle at a stop), while arcs represent either elementary connections (travel of a vehicle between consecutive stops), or transfer between different vehicles at the same stop, or waiting time between two time events (at the same stop). The arc weight is the time difference between the time events associated with the endpoints of the arc. Transfer times introduce realistic transfer restrictions between vehicles, and represent the required minimum time $transfer(S)$ that a passenger needs to be transferred between different vehicles within the same stop $S$.

A reduced version of the model (TE-red), eliminating nodes representing transfer events (without losing correctness) was also presented in [3,4,24].

In the rest of this section, we present the most recent graph-based approach for multimodal smart mobility in transist networks [14]. We follow the exposition in that paper and present first the dynamic timetable model (DTM), upon which the multimodal dynamic timetable model [14] is based.

**The Dynamic Timetable Model.** The *dynamic timetable model* (DTM) [3,4] aims at efficiently updating the timetable after a delay of a vehicle. Given a timetable $T = (Z, B, C)$, a directed graph $G = (V, E)$ representing DTM, is defined as follows: (1) for each stop $S$ in $B$, a *switch node* $\sigma_S$ is added to $V$, representing an arrival or start time event of a traveler at stop $S$; (2) for each elementary connection $c = (Z, S_d, S_a, t_d, t_a) \in C$ a *departure node* $d_c$ is added to $V$, and a *connection arc* $(d_c, \sigma_{S_a})$, connecting $d_c$ to the switch node $\sigma_{S_a}$ of (the immediately next stop) $S_a$, is added to $E$; (3) for each elementary connection $c = (Z, S_d, S_a, t_d, t_a) \in C$, a *switch arc* arc $(\sigma_{S_d}, d_c)$, connecting the switch node $s_{S_d}$ of the departure stop $S_d$ to the departure node $d_c$ of $c$ at $S_d$, is added to $E$; (4) for each vehicle $Z \in Z$ which travels through the itinerary $(c_1, c_2, \ldots, c_k)$, an arc (*vehicle arc*), connecting the departure node $d_{c_i}$ of $c_i$ with the departure node $d_{c_{i+1}}$ of $c_{i+1}$, is added to $E$, for each $i = 1, 2, \ldots, k - 1$.

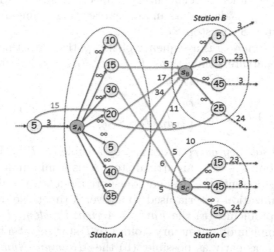

**Fig. 1.** A DTM graph. Switch nodes are drawn in blue. Departure nodes (yellow) are associated with the departure time of their corresponding elementary connection, and are ordered by arrival time at the (arrival) station. Switch arcs are drawn in brown, while vehicle arcs are drawn in green. (Color figure online)

The timetable routes are periodic, with period $T_p$ (typically $T_p = 1440$). Any transfer and travel time is assumed to last less than $T_p$. Given two time instances

$t_1$ and $t_2$, such that $t_1 \leq t_2$, $\Delta(t_1, t_2) = t_2 - t_1 (\text{mod } T_p)$ denotes the (cyclic) time difference between them. The time point $t(v) \in [0, T_p)$ of a departure node $v \in V$ is fixed and it denotes the scheduled departure time of the associated public transportation vehicle. The time point $t(v) \in [0, T_p)$ of a switch node $v \in V$ of stop $S \in \mathcal{B}$, varies and represents any possible start or arrival time at the stop $S$. The weight $w(e)$ of each non-switch (i.e., connection and vehicle) arc $e = (u, v) \in E$ is fixed and is set to $w(e) = \Delta(t(u), t(v))$. The weight of the switch arcs $e \in E$ varies and its default value is infinity.

For each connection $c = (Z, S_d, S_a, t_d, t_a)$, let $t_a(d_c)$ or $t_a(c)$ denote the arrival time $t_d + w(d_c, \sigma_{S_a})$ at stop $S_a$, departing from $S_d$ via the departure node $d_c$, at time $t(d_c) = t_d$. For each stop, the departure nodes are ordered by their $t_a(d_c)$ values (arrival times at their arrival stop $S_a$). For each switch node $\sigma_S$, its associated stop $S$ is stored, while for each departure node $d_c$, both the departure time reference $t_d(c)$ and the vehicle $Z(c)$ of connection $c$, with which $d_c$ is associated, are maintained. Figure 1 shows a DTM graph. Departures of station $A$ labeled 20 and 35 concern train connections, while the rest concern bus connections.

**The Multimodal Dynamic Timetable Model.** A *multimodal transport network* consists of schedule-based public transport along with road and pedestrian path networks, for supporting traveling with both unrestricted departure (e.g., for walking, cycling, and driving) and restricted departure (for embarking on public transport vehicles that follow scheduled timetables). In contrast to a restricted-departure timetable elementary connection, an unrestricted-departure connection $(\sigma_{S_A}, \sigma_{S_B})$ is defined as an arc representing a time-independent traveling path from stop $S_A$ to stop $S_B$.

A *multimodal itinerary* is a sequence of trip-paths consisting of unrestricted and restricted-departure connections $P = (c_1, c_2, \ldots, c_k)$ such that, for each $i = 2, 3, \ldots, k$, $S_a(c_{i-1}) = S_d(c_i)$ and

$$\Delta(t_a(c_{i-1}), t_d(c_i)) \geq \begin{cases} 0 & \text{if } Z(c_{i-1}) = Z(c_i) \\ transfer(S_a(c_{i-1})) & \text{otherwise.} \end{cases}$$

A *multimodal journey query* is defined by a tuple $(S, T, t_o, M)$ where $S \in \mathcal{B}$ is a departure stop, $T \in \mathcal{B}$ is an arrival stop, $t_o$ is a minimum departure time from $S$, and $M$ represents the desired transport mode(s). Recall that there are two natural optimization criteria used to answer a timetable query, giving rise to the following problems: (a) the *Earliest Arrival Problem (EAP)*, which asks for computing a multimodal itinerary from $S$ to $T$ starting at a time after $t_o$ and arriving at stop $T$ as early as possible; (b) the *Minimum Number of Transfers Problem (MNTP)*, which asks for computing a multimodal itinerary from $S$ to $T$ starting at a time after $t_o$ and having as few transfers from one vehicle to another as possible; and (c) the multicriteria version of EAP and MNTP (*multicriteria multimodal journey query*).

Given a timetable $\mathcal{T}$, a delay occurring on a connection $c$ is modelled as an increase of $\delta$ minutes on the arrival time: $t'_a(c) = t_a(c) + \delta (\text{mod } T_p)$. The timetable is then updated according to some specific policy which depends on the network infrastructure. The new disposition timetable $T'$, differs from $T$ oin the arrival and departure times of the vehicles that depend on $Z(c)$ in $T$. In [14] the simplest delay handling policy was considered. In particular, when a delay occurs on a connection $c$, the only time references which are updated are those regarding the departure times of $Z(c)$. Moreover, the policy does not take into account any possible slack times and hence the time references are updated by adding $\delta(\mathrm{mod}\ T_p)$.

The *multimodal dynamic timetable model* (MDTM) [14] is an extension of DTM, aiming at modeling traveling in multimodal transport networks. The key difference consists in the new ordering of the departure nodes within a stop.

Given a timetable $T = (\mathcal{Z}, \mathcal{B}, \mathcal{C})$, the directed graph $G = (V, E)$ representing MDTM is defined similarly to DTM, but with the following additional features:

- For each stop $S \in \mathcal{B}$, its associated departure nodes are grouped in a specific ordering: (i) a first grouping $\Gamma_1$ is created, where two departure nodes belong to the same group if the head switch node of their outgoing arcs is identical; (ii) within each group of $\Gamma_1$, a second grouping $\Gamma_2$ is created, where two departure nodes belong to the same group if the transport mode they represent is identical; and (iii) the departure nodes within each group of $\Gamma_2$ are ordered by increasing arrival time at the head switch nodes of their outgoing arcs.
- For each unrestricted-departure connection from stop $S_A$ to stop $S_B$, a switch-switch arc $(\sigma_{S_A}, \sigma_{S_B})$ is added to $G$.

Let $D_{S_d}(S_a, M)$ denote a group of departure nodes resulted from the aforementioned grouping, having departure stop $S_d$, arrival stop $S_a$, and transport mode $M$. Figure 2 shows the MDTM graph corresponding to the DTM graph of Fig. 1. Then, $D_{S_A}(S_B, bus)$ includes departure nodes 5 and 15 of stop $S_A$ that correspond to bus connections departing from $S_A$ and arriving at $S_B$.

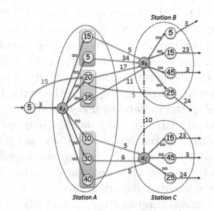

**Fig. 2.** The MDTM graph corresponding to the DTM graph of Fig. 1. Departure nodes grouping: light blue (brown) are train (bus) connections. The switch-switch arc (dotted black) introduces an unrestricted-departure connection between the stops. (Color figure online)

The algorithm for answering an EAP query $(S, T, t_s, M_{choices})$ resembles the execution of Dijkstra's algorithm on an MDTM graph $G$, starting from the switch node $\sigma_S$ of stop $S$. In particular, the query algorithm MDTM-QH maintains a set of *earliest arrival index tables* for each switch node of a stop. These tables contribute to the effective searching of the departure nodes that have both valid departure times from some stop $S_d$ and earliest arrival times to an adjacent arrival stop $S_a$. Such an earliest arrival index table $I_{S_d}(S_a, M)$, consisting of departure nodes with departure stop $S_d$, arrival stop $S_a$ and transport mode $M$, is constructed as follows: Let $d_1, d_2, .., d_k \in D_{S_d}(S_a, M)$ be the sequence of the departure nodes, ordered by arrival time at $S_a$, for a trip departing from stop $S_d$ and arriving to stop $S_a$ with transport mode $M$. Initially, $I_{S_d}(S_a, M)$ is empty. Node $d_1$ is inserted in $I_{S_d}(S_a, M)$ and $t_{max} = t(d_1)$ is the current departure time. Then, for $i = 2, ..., k$, if $t_{max} < t(d_i)$, then $t_{max} = t(d_i)$ and $d_i$ is inserted at the end of $I_{S_d}(S_a, M)$ table; otherwise, $d_i$ is skipped. If the table contains the departure nodes $v_1, ..., v_l$, $l \leq k$, then, if $t(\sigma_{S_d}) \in [t(v_i), t(v_{i+1}))$ then $v_i$ is the first departure node to start the search of the earliest arrival time at $S_a$. This means that the earliest arrival index allows us to bypass all of the departure nodes with departure time less than $t(v_i)$. Also for the endpoints, if $t(\sigma_{S_d}) \geq t(v_l)$ $(t(\sigma_{S_d}) \leq t(v_1))$ then $v_l$ $(v_1)$ is the first departure node to get the earliest arrival time at $S_a$.

To reduce the size and the operations in the priority queue of the query algorithm, only the switch nodes are inserted in it, and the arc relaxation is modified as described below (iteration step).

Overall, the MDTM-QH algorithm works as follows.

- *Initialization.* The switch node $\sigma_{S_o}$ of the origin stop $S_o$ is inserted in the priority queue, with distance $dist[\sigma_{S_o}] = t_s$ and time $t(\sigma_{S_o}) = t_s$. During the algorithm execution, provided that the traveller is already at $S_o$ at time $t_s$ the minimum transfer time of $S_o$ is set to $transfer(S_o) = 0$.
- *Iteration.* At each step, a switch node $\sigma_{S_x}$ is extracted from the priority queue. The node $\sigma_{S_x}$ is settled having got the earliest arrival time for the optimal journey departing from $S_o$ at time $t_s$ and arriving to $\sigma_{S_x}$ at time $dist[\sigma_{S_x}]$. Then the algorithm relaxes the outgoing arcs of $\sigma_{S_x}$ following a node filtering and blocking process: (i) using the existing grouping of departure nodes ($\Gamma_1$ and $\Gamma_2$), the departure node groups corresponding to non-selected transportation modes are skipped; (ii) using the earliest arrival index tables, the algorithm can skip the departure nodes of stop $S_x$ that have an earlier than $dist[\sigma_{S_x}]$ departure time or they provide non-optimal arrival times to the next adjacent stops. In particular, after $\sigma_{S_x}$ is extracted, then for each adjacent arrival stop $S_a$ of stop $S_x$ and for each enabled transport mode $M \in M_{choices}$: (1) a binary search is performed on the index table $I_{S_x}(S_a, M)$ for getting the first contented departure node $d_r$ with $t(d_r) > t(\sigma_{S_x})$ that provide the earliest arrival time at stop $S_a$; and (2) an arc relaxation step is performed based on those departure nodes.
- *Termination.* The algorithm terminates when the switch node $\sigma_{S_d}$ of the destination stop $S_d$ is settled (extracted from the priority queue).

Let the sequence of the outgoing switch arcs of $\sigma_{S_x}$ be $e_1, e_2, ..., e_{r-1}, e_r, ..., e_k$, which corresponds to a travel using the transport mode $M$ and arriving at the stop $S_a$. Let $e_i = (\sigma_{S_x}, d_i)$, $i = 1, ...k$, and let the node $d_r$ be the departure node that is returned by the $I_{S_x}(S_a, M)$. Within the current time period $T_p$, arcs $e_1, e_2, ..., e_r$ can be safely skipped, because they provide earlier departures or non-optimal arrival paths from $S_x$ to $S_a$. The first arc that is relaxed is $e_r$. Provided that the next switch arcs and their departure node heads are ordered by arrival time, the algorithm relaxes the arcs $e_r, ..., e_k, e_1, ..., e_{r-1}$ and it stops as soon as falls over a departure node $d_i$ with (a) $\Delta(t(\sigma_{S_x}), td(d_i)) > transfer(S_x)$ and (b) $\Delta(td(d_i), t(\sigma_{S_x}) + w(d_i, \sigma_{S_a}) > dist[\sigma_{S_a}] + transfer(\sigma_{S_a})$. The first condition ensures the minimum transfer time for the traveller on using a different vehicle to continue its travel. The second condition checks when the next departure nodes with different vehicles cannot provide better arrival times.

At each case, if the departure node head $d_i$ of the switch arc $e_i$ is visited for the first time or it has from a previous step a greater distance, then $dist[d_i] = dist[\sigma_{S_x}] + \Delta(t(\sigma_{S_x}, t_d(d_i))$ and $w(\sigma_{S_x}, d_i) = dist[d_i] - dist[\sigma_{S_x}]$. When the distance is updated, the algorithm relaxes also the outgoing arcs of the departure node $d_i$. For its outgoing arc $(d_i, \sigma_{S_a})$, if the $\sigma_{S_a}$ is first time visited or has from a previous step a greater distance, then $dist[\sigma_{S_a}] = dist[d_i] + w(d_i, \sigma_{S_a})$. Also if there is a vehicle (departure-departure) arc $(d_i, d_j)$, then for the associated vehicle the outgoing arcs of the departure nodes $d_j, ..., d_{last}$ are also relaxed at the next stops at which the vehicle passes, for the same route. This process stops if it falls over a departure which has an outgoing arc to a switch node which is not yet visited or has not yet been extracted from the priority queue.

The performance of the MDTM-QH algorithm can be enhanced by combining it with the ALT heuristic [15], resulting in algorithm MDTM-QH-ALT. Since the choice of good landmarks is crucial when using ALT, the approach in [14] follows those in [3,4,7] and selects as landmarks the switch nodes. Algorithm MDTM-QH-ALT reduces considerably the search space and boosts query performance.

Answering multimodal multicriteria queries on the EA and MNT criteria, results in computing an exponential in size set of Pareto-optimal journeys. For this reason, the focus in [14] is on finding a solution that minimizes MNT, while retaining the EA below a given threshold $P$ (a variant also considered in [24]), resulting on the multicriteria algorithms McMDTM-QH and McMDTM-QH-ALT (combination with ALT).

McMDTM-QH works as follows. Let $(S, T, t_s, M_{choices})$ be a multicriteria (EA, MNT) query, starting from the switch node $\sigma_S$ of stop $S$. The number of transfers is taken into account by setting the weight of all switch-departure arcs to 1 (representing a transfer between vehicles) and the weight of the rest of the arcs to 0. Due to the modeling, every single switch node in MDTM can have at least as many Pareto-optimal solutions as its incoming arcs. Initially, the cost minimization is on EA. Therefore, when the target switch node is settled, the first (EA, MNT) Pareto optimal journey has been found with the minimum arrival time $A_{min}$. Then Dijkstra's algorithm continues and whenever the target switch node is explored again with a smaller number of transfers than in any of the already found Pareto-optimal solutions, a new Pareto-optimal journey is found. The algorithm stops when all journey solutions, with arrival time to the target stop less or equal than $P \cdot A_{min}$, have been found.

For updating the timetable when a delay occurs, algorithm MDTM-U is called to update the corresponding MDTM graph. Given a timetable $\mathcal{T}$, assume that a delay $\delta$ occurs first in a connection $c_0$ of $\mathcal{T}$, and it is propagated to the (affected) connections $c_0, c_1, ..., c_k$, which are performed by the same vehicle. Also let $d_0, d_1, ..., d_k$ be the departure nodes corresponding to the affected connections. If $\mathcal{T}$ is represented as an MDTM graph $G$, then MDTM computes the MDTM graph $G'$, corresponding to the disposition timetable $\mathcal{T}'$, as follows.

- Edge weight increase: Starting with $c_0 = (Z, S_d, S_a, t_d, t_a)$, the weight of arc $(d_{t_d}, \sigma_{S_a})$ is increased by $\delta$.
- Node reordering: For each of the other connections $c_i, i = 1, .., k$, its associated departure node $d_i$ has its departure time $t_d(d_i)$ increased by $\delta$. Due to that increase, the arrival time ordering of the departure nodes on the affected stops may be invalidated. Hence, along with the new arrival times, the departure node $d_i$ might need to be moved to its correct position within its group, i.e., before a departure node with arrival time greater than $t_a(d_i)$.

### 3.3    Practical Performance

The aforementioned multimodal journey planning algorithmic approaches of RAPTOR, McRAPTOR, and DTM constituted the core algorithmic routing engines of a cloud-based mobility platform developed in the frame of [20] that delivers personalized mobility services in smart cities [11]. These services have been evaluated during a pilot study carried out in real-world conditions in the city of Vitoria-Gasteiz. Moreover, the mobile client application of the cloud-based journey planner based on DTM has been enhanced with an additional user assessment feature that allows users to assess the suggested itineraries offered by the application [13].

In this section, we further report on the practical performance of the afore-mentioned approaches also in comparison with the MDTM model on real-world data sets carried out in [14].

The data sets consisted of the metropolitan public transit networks of Berlin and London, taken from [26] and [27], respectively, and which were integrated with their corresponding road and pedestrian networks taken from [23]. The packed-memory graph structure [21] was used for representing the input graph instances.

The metropolitan public transit network of Berlin resulted in an MDTM graph of about 4.3 Million nodes and 12.7 Million arcs, with average transfer time 0.7 min, average degree of adjacent stops/stations 2.7, and distribution of transportation means 76% bus, 15% train and 9% tram.

The metropolitan public transit network of London resulted in an MDTM graph of about 14 Million nodes and 41.8 Million arcs, with average transfer time 0.8 min, average degree of adjacent stops/stations 1.2, and distribution of transportation means 98% bus, and 2% train.

For assessing the practical performance of MDTM, non-restricted departure traveling paths were additionally added in [14], using two approaches:

– Limited walking and driving travel time paths on transitively closed pedestrian and road networks. Via the pedestrian networks, single foot-paths for enabling walking between nearby stops were added having a shortest travel time of at most 10mins, with walking speed 1 m/sec. Also, via the road networks, free flow speed driving-paths were added for enabling the driving between stops with EVs, considering for this scenario 10 EV-stations providing public communal EVs with shortest travel time of at most 1 h. Driving-paths connect only EV-stations. In the Berlin instance, the switch-switch arcs representing foot-paths are 2381 and the driving-paths are 39. In the London instance, the switch-switch arcs representing foot-paths are 37226 and the driving-paths are 60.
– Unlimited walking travel time paths on the full pedestrian network. For this purpose, each switch node in the public transit network was connected with the nearest node in the pedestrian network by an access edge (an approach inspired by that in [28]). In the Berlin instance, the embedded pedestrian network had 932108 nodes and 1059556 edges. In the London instance, the embedded pedestrian network had 1520056 nodes and 1653052 edges.

The practical assessment in [14] on the aforementioned data sets concerned the generation, for each instance, of 10K random queries consisting of source and target stop pairs, along with a departure time at each source stop. In the evaluation, the following EA query algorithms were included: TE-QH-ALT for TE-red [4], DTM-QH-ALT for DTM [4], and MDTM-QH-ALT for MDTM [14]. For the latter, the multicriteria (EA, MNT) query algorithm McMDTM-QH-ALT with threshold $P$ on EA 100% and 120% (denoted with extension 1.0 and 1.2, respectively) were also included. The results of the algorithms for answering multimodal queries are reported in Table 1.

We added in Table 1 the query times of the best previous (*RAPTOR* based) approaches MCR-ht and MR-∞-t10[2] in [5,9]. Note that the former computes multicriteria (on arrival time, number of transfers, and walking duration) multimodal journeys, while the latter computes multicriteria (on arrival time and number of transfers) multimodal journeys. The times are scaled versions of those reported in [5,9] using the benchmark for scaling factors in [25]. Since MCR-ht, MR-∞-t10, McMDTM-QH-ALT-1.0 and McMDTM-QH-ALT-1.2 report multicriteria multimodal queries, it is natural that they take more time than regular multimodal EAP (unicriterion) query algorithms. This is also true for the case of unlimited walking, due to the much larger search space explored by the algorithms. Nevertheless, McMDTM-QH-ALT-1.0 and McMDTM-QH-ALT-1.2 are competitive to MCR-ht and MR-∞-t10.

---

[2] MCR-ht weakens the domination rules by trading off walking and arrival time. In MR-∞-t10 the walking duration is not used as criterion and it is limited to 10 min.

**Table 1.** Comparison between query algorithms. L-Walk (U-Walk) denotes a query algorithm with limited (unlimited) walking. Bullets (•) indicate the options taken into account. MC denotes a multicriteria journey on arrival time and number of transfers (and walking duration for MCR-ht).

|  | Algorithm | MC | Travel modes | | | | | Query [ms] | |
|---|---|---|---|---|---|---|---|---|---|
|  |  |  | Bus | Train | Walk | EV/Car | Cycle | L-Walk | U-Walk |
| Berlin | TE-QH-ALT [4] |  | • | • |  |  |  | 6.88 |  |
|  | DTM-QH-ALT [4] |  | • | • |  |  |  | 12.17 |  |
|  | MDTM-QH-ALT [14] |  | • | • |  |  |  | 6.12 |  |
|  | MDTM-QH-ALT [14] |  | • | • | • | • |  | 8.49 | 105.12 |
| London | TE-QH-ALT [4] |  | • | • |  |  |  | 5.14 |  |
|  | DTM-QH-ALT [4] |  | • | • |  |  |  | 10.25 |  |
|  | MDTM-QH-ALT [14] |  | • | • |  |  |  | 4.17 |  |
|  | MDTM-QH-ALT |  | • | • | • | • |  | 6.10 | 114.88 |
|  | McMDTM-QH-ALT-1.0 [14] | • | • | • | • | • |  | 6.29 | 216.36 |
|  | McMDTM-QH-ALT-1.2 [14] | • | • | • | • | • |  | 15.44 | 360.94 |
|  | MCR-ht [5,9] | • | • | • | • |  | • |  | 361.23 |
|  | MR-∞-t10 [5,9] | • | • | • | • |  | • | 21.47 |  |

For evaluating updates (occurring after a delay), 1000 elementary connections were randomly selected, for each input instance, and for each elementary connection a delay was randomly generated, affecting the corresponding train or bus, chosen with uniform probability distribution between 1 and 360 min.

**Table 2.** Comparison among update algorithms.

| Instance | Algorithm | Travel modes | | Update [$\mu$s] |
|---|---|---|---|---|
|  |  | Bus | Train |  |
| Berlin | TE-UH [4] | • | • | 238.5 |
|  | DTM-U [4] | • | • | 80.2 |
|  | MDTM-U [14] | • | • | 84.4 |
| London | TE-UH [4] | • | • | 477.2 |
|  | DTM-U [4] | • | • | 122.8 |
|  | MDTM-U [14] | • | • | 137.5 |

In the experimental evaluation we have included the update algorithms TE-UH for TE-red [4], DTM-U for DTM [4], and MDTM-U for MDTM [14]. The experimental results of the update algorithms are reported in Table 2. The updates times measure the average computational times for updating the graphs when a delay in a transportation vehicle itinerary has to be absorbed.

# 4    Conclusions

We provided recent innovative algorithmic technology required by core routing engines of cloud-based applications for mobility in urban environments. Most of the aforementioned algorithmic technology has been embedded in the integrated mobility application for smart cities developed in the frame of [20], whose back-end services, their interrelation and rationale can be found in [11].

# References

1. Bast, H., et al.: Route planning in transportation networks. In: Kliemann, L., Sanders, P. (eds.) Algorithm Engineering. LNCS, vol. 9220, pp. 19–80. Springer, Cham (2016). https://doi.org/10.1007/978-3-319-49487-6_2
2. Baum, M., Dibbelt, J., Pajor, T., Wagner, D.: Dynamic time-dependent route planning in road networks with user preferences. In: Goldberg, A.V., Kulikov, A.S. (eds.) SEA 2016. LNCS, vol. 9685, pp. 33–49. Springer, Cham (2016). https://doi.org/10.1007/978-3-319-38851-9_3
3. Cionini, A., et al.: Engineering graph-based models for dynamic timetable information systems. In: 14th Workshop on Algorithmic Approaches for Transportation Modelling, Optimization, and Systems (ATMOS2014). OASICS, vol. 42, pp. 46–61. Schloss Dagstuhl (2014)
4. Cionini, A., et al.: Engineering graph-based models for dynamic timetable information systems. J. Discret. Algorithms **46–47**, 40–58 (2017)
5. Delling, D., Dibbelt, J., Pajor, T., Wagner, D., Werneck, R.F.: Computing multimodal journeys in practice. In: Bonifaci, V., Demetrescu, C., Marchetti-Spaccamela, A. (eds.) SEA 2013. LNCS, vol. 7933, pp. 260–271. Springer, Heidelberg (2013). https://doi.org/10.1007/978-3-642-38527-8_24
6. Delling, D., Goldberg, A.V., Pajor, T., Werneck, R.F.: Customizable route planning in road networks. Transp. Sci. **51**(2), 566–591 (2015)
7. Delling, D., Pajor, T., Wagner, D.: Engineering time-expanded graphs for faster timetable information. In: Ahuja, R.K., Möhring, R.H., Zaroliagis, C.D. (eds.) Robust and Online Large-Scale Optimization. LNCS, vol. 5868, pp. 182–206. Springer, Heidelberg (2009). https://doi.org/10.1007/978-3-642-05465-5_7
8. Delling, D., Pajor, T., Werneck, R.F.: Round-based public transit routing. Transp. Sci. **49**(3), 591–604 (2015)
9. Dibbelt, J.: Engineering algorithms for route planning in multimodal transportation networks. Ph.D. thesis, Karlsruhe Institute of Technology, February 2016
10. Dreyfus, S.E.: An appraisal of some shortest-path algorithms. Oper. Res. **17**(3), 395–412 (1969)
11. Gavalas, D., et al.: Renewable mobility in smart cities: cloud-based services. In: Proceedings of 23rd IEEE Symposium on Computers and Communications – ISCC 2018. IEEE Computer Society (2018, to appear)
12. Giannakopoulou, K., Kontogiannis, S., Papastavrou, G., Zaroliagis, C.: A cloud-based time-dependent routing service. In: Sellis, T., Oikonomou, K. (eds.) ALGO-CLOUD 2016. LNCS, vol. 10230, pp. 41–64. Springer, Cham (2017). https://doi.org/10.1007/978-3-319-57045-7_4
13. Giannakopoulou, K., Nikoletseas, S., Paraskevopoulos, A., Zaroliagis, C.: Dynamic timetable information in smart cities. In: Proceedings of 22nd IEEE Symposium on Computers and Communications – ISCC 2017, pp. 42–47. IEEE Computer Society (2017)

14. Giannakopoulou, K., Paraskevopoulos, A., Zaroliagis, C.: Multimodal dynamic journey planning. In: Proceedings of 23rd IEEE Symposium on Computers and Communications – ISCC 2018. IEEE Computer Society (2018, to appear)
15. Goldberg, A., Harrelson, C.: Computing the shortest path: A* search meets graph theory. In: ACM-SIAM Symposium on Discrete Algorithms (SODA 2005), pp. 156–165. SIAM (2005)
16. Kontogiannis, S., Michalopoulos, G., Papastavrou, G., Paraskevopoulos, A., Wagner, D., Zaroliagis, C.: Engineering oracles for time-dependent road networks. In: Algorithm Engineering and Experiments – ALENEX 2016, pp. 1–14. SIAM (2016)
17. Kontogiannis, S., Papastavrou, G., Paraskevopoulos, A., Wagner, D., Zaroliagis, C.: Improved oracles for time-dependent road networks. In: Algorithmic Approaches for Transportation Modeling, Optimization, and Systems - ATMOS 2017. OASIcs, vol. 59, pp. 4:1–4:17 (2017)
18. Kontogiannis, S., Wagner, D., Zaroliagis, C.: Hierarchical time-dependent oracles. In: Algorithms and Computation – ISAAC 2016. LIPIcs, vol. 64, pp. 47:1–47:13 (2016)
19. Kontogiannis, S., Zaroliagis, C.: Distance oracles for time-dependent networks. Algorithmica **74**(4), 1404–1434 (2016)
20. MOVESMART EU FP7 project. https://cordis.europa.eu/project/rcn/110310_en.html
21. Mali, G., Michail, P., Paraskevopoulos, A., Zaroliagis, C.: A new dynamic graph structure for large-scale transportation networks. In: Spirakis, P.G., Serna, M. (eds.) CIAC 2013. LNCS, vol. 7878, pp. 312–323. Springer, Heidelberg (2013). https://doi.org/10.1007/978-3-642-38233-8_26
22. Müller-Hannemann, M., Schulz, F., Wagner, D., Zaroliagis, C.: Timetable information: models and algorithms. In: Geraets, F., Kroon, L., Schoebel, A., Wagner, D., Zaroliagis, C.D. (eds.) Algorithmic Methods for Railway Optimization. LNCS, vol. 4359, pp. 67–90. Springer, Heidelberg (2007). https://doi.org/10.1007/978-3-540-74247-0_3
23. OpenStreetMap Data Extracts. http://download.geofabrik.de
24. Pyrga, E., Schulz, F., Wagner, D., Zaroliagis, C.: Efficient models for timetable information in public transportation systems. ACM J. Exp. Algorithmics **12**(2.4), 1–39 (2008)
25. Reference CPU scores. http://i11www.iti.kit.edu/~pajor/survey
26. Transit Feeds. https://transitfeeds.com
27. Transport for London. https://tfl.gov.uk
28. Wagner, D., Zündorf, T.: Public transit routing with unrestricted walking. In: Algorithmic Approaches for Transportation Modeling, Optimization, and Systems – ATMOS 2017. OASIcs, vol. 59, pp. 7:1–7:14 (2017)

# A Frequent Itemset Hiding Toolbox

Aris Gkoulalas-Divanis[1]([⊠]), Vasileios Kagklis[2], and Elias C. Stavropoulos[3]

[1] IBM Watson Health, Cambridge, MA, USA
gkoulala@us.ibm.com
[2] Computer Engineering and Informatics Department, University of Patras,
Patras, Greece
kagklis@ceid.upatras.gr
[3] Educational Content, Methodology and Technology Laboratory,
Hellenic Open University, Patras, Greece
estavop@eap.gr

**Abstract.** Advances in data collection and storage technologies have
given way to the establishment of transactional databases among companies and organizations, as they allow enormous amounts of data to be
stored efficiently. Useful knowledge can be mined from these data, which
can be used in several ways depending on the nature of the data. Quite
often companies and organizations are willing to share data for the sake
of mutual benefit. However, the sharing of such data comes with risks,
as problems with privacy may arise. Sensitive data, along with sensitive
knowledge inferred from this data, must be protected from unintentional
exposure to unauthorized parties. One form of the inferred knowledge
is frequent patterns mined in the form of frequent itemsets from transactional databases. The problem of protecting such patterns from being
discovered, is known as the frequent itemset hiding problem. In this paper
we present a toolbox, which provides several implementations of frequent
itemset hiding algorithms. Firstly, we summarize the most important
aspects of each algorithm. We then introduce the architecture of the
toolbox and its novel features. Finally, we provide experimental results
on real world datasets, demonstrating the capabilities of the toolbox and
the convenience it offers in comparing different algorithms.

**Keywords:** Privacy preserving data mining · Knowledge hiding ·
Frequent itemset hiding · Sensitive knowledge

## 1 Introduction

Nowadays, transactional databases are being used more and more by organizations, as they support efficient storage of large volumes of data. By using data
mining techniques on such data, modern companies can extract useful information that can help these companies understand the behavior of their customers,
support decision making, plan their business strategy, etc.

Companies and organizations are willing to share data for the sake of mutual
benefit. The benefits derived from the sharing of such data are considerable

© Springer Nature Switzerland AG 2019
Y. Disser and V. S. Verykios (Eds.): ALGOCLOUD 2018, LNCS 11409, pp. 169–182, 2019.
https://doi.org/10.1007/978-3-030-19759-9_11

and they cannot be ignored. A typical example is a supermarket, which collects market basket data of its customers' purchases on a regular basis. These organizations might be willing to share their collected information with other parties, such as advisory organizations, for mutual benefit. For example. two stores, say A and B, cooperate in order to discover their customers' purchase behaviors.

Unfortunately, the sharing of such data does not come without risks, as problems with privacy may arise. Therefore, it must be done in such a way, that no sensitive information will be exposed to unauthorized parties. In our previous example, there might be some sensitive information that could reflect the business strategies and secrets of the participating companies that should not be revealed to their adversary competitors. For example, if data analysts of store A found out that its customers tend to purchase products x and y at the same time, they should regard this knowledge as sensitive information, and not disclose it to store B. With this knowledge, store B could offer sales with a lower price for customers who buy x and y together. Then, store A could possibly face the danger of losing some of its customers. Verykios et al. [33], Oliveira and Zaïane [25], and Evfimievski et al. [15] discuss other examples of situations where the sharing of operational databases could have serious adverse effects.

Privacy preserving data mining (PPDM) [3,22] is the research area that investigates techniques to preserve the privacy of data and patterns. Knowledge hiding [19], which is a subfield of PPDM, has as its goal to prevent the exposure of sensitive patterns included in the data to be published. Knowledge hiding can be achieved in several ways. The most commonly used is through the sanitization [4] of a number of transactions in the database, so that the sensitive information can no longer be extracted. Therefore, a hiding technique must be applied before making a database available for sharing. Many data mining tasks rely on frequent itemsets to be identified as a first step in their process. Thus, concealing the frequent patterns associated with the sensitive information would guarantee the preservation of the privacy of the sensitive relationships between patterns of the itemsets that may be discovered through any of these data mining tasks.

In this paper, we present the software architecture and implementation of a frequent itemset hiding (FIH) toolbox, which can be used to apply a suite of hiding techniques on real world datasets. The toolbox comes with a built-in library containing several implementations of FIH algorithms and a suite of performance metrics. Lastly, we present experimental results, to demonstrate the efficiency of the toolbox and the convenience it offers to data owners in comparing different frequent itemset hiding algorithms.

The rest of this paper is organized as follows. Section 2 provides an overview of the related work. In Sect. 3 we present the necessary background information and define the FIH problem. Section 4 describes the software architecture of the FIH toolbox and Sect. 5 presents its features. Section 6 summarizes the evaluation process and presents the experimental results. Finally, Sect. 7 concludes this work.

## 2   Related Work

Clifton et al. [12,13] are among the first to deal with privacy preservation in the field of data mining and propose data-obscuring techniques in order to avoid discovery of sensitive patterns. Atallah et al. [5] prove that optimally solving the frequent itemset hiding problem is NP-hard. The authors also present a greedy algorithm that turns ones into zeros in the database in order to hide sensitive frequent patterns. Various extensions of this work have been proposed over the years, including those by Verykios et al. [33] and Dasseni et al. [14].

Saygin et al. [30] and Verykios et al. [34] present another approach, according to which items may be added or removed from transactions, recording the possible participation of certain items to transactions with "?" (question marks), so that the hiding is achieved not by falsifying, but by fuzzifying the data. Oliveira and Zaïane [26,27] propose a technique for hiding multiple association rules simultaneously that requires only one pass over the whole database, regardless the size of the database. Pontikakis et al. [28,29] perform an exhaustive evaluation of distortion and blocking (using question marks for hiding) techniques. Bertino et al. [8] propose an evaluation framework that aims at measuring the performance of frequent itemset and association rule hiding techniques.

Menon et al. [23] were the first to introduce an integer linear program (ILP) formulation of the frequent itemset hiding problem. The solution of their ILP points out which transactions need to be sanitized in order to conceal the sensitive patterns. The sanitization process is addressed as a separate phase, independently of the linear programming solution, in a heuristic yet suboptimal way. Sun and Yu [32] introduce a greedy border-based approach for hiding sensitive frequent itemsets. They propose an algorithm that takes advantage of the interplay between preserving maximally non-sensitive and downsizing minimally sensitive frequent itemsets, and gives an accurate and efficient hiding solution.

In [20], Kagklis et al. formulate the FIH problem as an ILP and present a heuristic approach to calculate the coefficients of the objective function of the ILP, while at the same time minimizing the side effects introduced by the hiding process. They also propose a sanitization algorithm for the hiding process.

Stavropoulos et al. [31] relied on the enumeration on the minimal transversals of a hypergraph in order to induce the ideal border between frequent and sensitive itemsets. The ideal border is then utilized to formulate an ILP, the solution of which identifies the set of transactions that need to be sanitized so that the hiding can be achieved with maximum accuracy.

A large number of different approaches towards solving the frequent itemset hiding problem have been proposed. Additionally, several performance evaluation metrics or frameworks have been developed, so as to compare the quality of these techniques. Nevertheless, to our knowledge, there is yet no publicly available tool that offers implementations of such techniques and/or evaluation metrics, along with a common ground for performing experimental evaluation. The work proposed in this paper presents such a toolbox, which can be easily extended to host other hiding techniques, as well as additional evaluation metrics and visualization capabilities.

# 3    Problem Formulation

## 3.1    Definitions

Let $I = \{i_1, i_2, ..., i_n\}$ be a set of items. Any non empty subset of $I$, $X \subseteq I$, is called an itemset. An itemset consisting of $k$ elements is called a $k$-itemset. A transaction $T$ is a pair $T = (tid, Z)$, where $Z \subseteq I$ is the itemset and $tid$ is a unique identifier, used to distinguish among transactions that correspond to the same itemset. Let also $D$ be a set of transactions. A transaction $T$ supports an itemset $X$, if and only if $X \subseteq T$. Given a set of items $I$, we will denote as $P(I)$ the powerset of $I$, that is all possible combinations of items from $I$.

The support of an itemset $X$ in a database $D$, denoted as $\sigma_D(X)$, is the number of transactions containing $X$. Support can also be expressed as the percentage of transactions in $D$. An itemset $X$ is frequent in $D$, if and only if its support in $D$ is at least equal to a support threshold $\sigma_{min}$. The set of all frequent itemsets, denoted as $F$, is given by $F = \{X \subseteq I \mid \sigma_D(X) \geq \sigma_{min}\}$.

Let $S$ be the set of sensitive (frequent) itemsets that need to be hidden. We will denote as $I^S$ the set of different items contained in $S$. Moreover, let all sensitive itemsets and their supersets in $F$ be denoted as $SS$, where $SS = \{X \in F \mid \forall Y \in S, X \supseteq Y\}$ and $S \subseteq SS \subseteq F$. The revised set of frequent itemsets, denoted as $\widetilde{F}$, is given by $\widetilde{F} = F - SS$.

By its definition, the revised set of frequent itemsets $\widetilde{F}$ is the ideal set of itemsets that would still remain frequent after hiding the sensitive itemsets. The ideal case is when only the sensitive itemsets and their supersets are concealed. Sensitive itemsets is desirable to get concealed, while their supersets are inevitably concealed, due to the antimonotonicity property of support, that is $\forall X, Y \subseteq I$: $X \subseteq Y \Rightarrow \sigma_D(X) \geq \sigma_D(Y)$.

## 3.2    Problem Description

We are given a transactional database $D$, a support threshold $\sigma_{min}$ and a set of sensitive frequent itemsets $S$. The frequent itemset hiding (FIH) problem involves sanitizing selected transactions in database $D$, so that itemsets in $S$ cannot be mined from the sanitized database $D'$ using a support threshold equal or above $\sigma_{min}$. A database is said to be sanitized, if it is altered in such a way that it no longer supports any sensitive itemset in it. Sanitizing a database involves the sanitization of one or more transactions. A transaction is said to be sanitized, if it is altered so that it no longer supports any sensitive itemset. In the best case scenario, the sensitive itemsets should be hidden with minimal damage to the database. In other words, we want to conceal the itemsets in $S$, whilst having the minimum impact on the utility of the database.

# 4    The Architecture of the FIH Toolbox

From an abstract point of view, the toolbox is divided into four layers (see Fig. 1). The top layer is the "Presentation Layer", which implements the Graphical User

Interface (GUI) of the toolbox and provides visualizations of the results received from the next layer below. Users interact with the toolbox through the presentation layer, depending on the options they make. The GUI is implemented in Python, using the Tkinter and ttk modules.

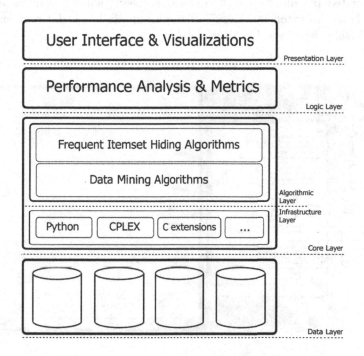

**Fig. 1.** The architecture of the FIH toolbox.

The next layer is the "Logic Layer". In this layer, the performance analysis phase is implemented and the calculation of the metrics take place. Such metrics are the number of changes made in the raw data, the number of side effects, the execution time and the information loss incurred. A detailed description of the performance metrics is presented in Sect. 6. This layer is also responsible for moving and processing data between its two surrounding layers.

The third layer is the "Core Layer", which consists of two sublayers: the "Algorithmic Sublayer" and the "Infrastructure Sublayer". The "Algorithmic Sublayer" is basically the built-in library with the implemented FIH algorithms, described in Sect. 5.1, along with the implementation of some basic data mining methods. Currently, the only data mining method offered is the Apriori algorithm [3]. The "Infrastructure Sublayer" consists of all the tools used for the development of the toolbox. Most of the code is implemented in Python. The hiding algorithms are implemented in Cython [7]. The Apriori algorithm [3] is part of the PyFIM extension module. This module is a C extension for Python [10] to efficiently mine the set of frequent itemsets. The IBM ILOG CPLEX 12.6 [2] is used for solving any linear program.

Finally, the last layer is the "Data Layer", which is related to the datasets that can be used and their corresponding supported formats. The toolbox supports both space-separated value (.ssv) and comma-separated value (.csv) formatted files. The delimiter is declared by the extension of the file. Therefore, a space-separated file must have the extension ".ssv", while a comma-separated file must have the extension ".csv". If the input file has a different extension, then by default the space delimiter is used.

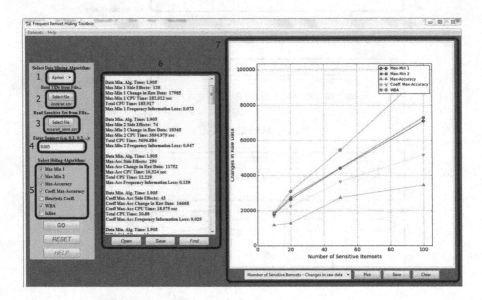

**Fig. 2.** The GUI of the FIH toolbox.

In Fig. 2, the GUI of the toolbox is presented. For the time being, there is a beta version of the toolbox available [1]. Linear programming techniques require a license for CPLEX, which can be obtained for free through IBM's Academic Initiative program. The user can apply a hiding technique on a dataset by following a few easy steps, which are summarized in Fig. 3.

Firstly, a data mining algorithm from *field 1* must be selected. Then, a dataset must be supplied by the user, by using *field 2*. Respectively, the file with the sensitive itemsets must be given by using *field 3*. In *field 4*, the support threshold must be specified. *Field 5* is a group of checkboxes. By checking a checkbox, the user selects the corresponding algorithm to be executed. *Field 6* is a text editor, where the sanitized dataset and the calculated metrics are printed. The user can save these results by using the "Save" button below the text editor. Finally, *field 7* is a canvas that displays visualizations of the metrics. Below the canvas, there is drop-down list with options related to the axes of the figures. The "Plot" button should be used after an option from the drop-down list is selected, so as to plot the corresponding figure. Buttons "Save" and "Clear" can be used to save and clear the current figure respectively.

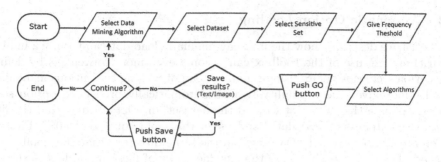

**Fig. 3.** "How-to-use" flow chart for the FIH toolbox.

# 5 Special Features of the FIH Toolbox

## 5.1 Built-In Library

The toolbox comes with a built-in library that consists of the following FIH algorithm implementations as independent modules: Max-Min Algorithms [24], Weight-Based Approach [32], Max-Accuracy Algorithm [23], Coefficient-Based Max-Accuracy Algorithm [21], Heuristic Coefficient Based Approach [20], and Inline Algorithm [16]. The Max-Min algorithms (Max-Min 1, Max-Min 2) and the Weight-Based Approach (WBA) use the border revision theory [32]. The Max-Accuracy algorithm, the Coefficient-Based Max-Accuracy algorithm and the Heuristic Coefficient Based Approach formulate the problem as an ILP. A heuristic algorithm for the sanitization is also used. The Inline algorithm combines border revision theory and linear programming. Future versions of the tool will include algorithms in [17,18], as well as methods for hiding sensitive association rules [5, 14, 25, 26, 28–30, 33, 34].

## 5.2 Extensibility

The toolbox comes with a built-in library, which contains some implemented FIH algorithms. However, a non-extensible library would limit the utility of the Toolbox. An important feature of the toolbox is that it can be extended by its users. Users can implement and import new algorithms, and compare them with the existing algorithms in the built-in library.

User-implemented algorithms must be compatible with the toolbox. Therefore, any implementation must comply to the restrictions and guidelines, as described in the manual that can be found in [1]. Users can easily implement compatible source files by following the instructions given in the manual. After the implementation is completed, it can be used right away; create a folder with the name "Extensions" (without the quotes) in the same directory with the Toolbox and copy the source file in it.

## 5.3  Automatic Option Loading

In Sect. 4, we described how the user can manually load data and apply a hiding algorithm. The use of the toolbox can become even more convenient by defining option scenarios before using them. Instead of making the options manually, the Toolbox gives the capability to load automatically predefined option scenarios. Assume that we want to load the dataset "myDataset.dat" and the file of sensitive itemsets "HS1.dat", and use a threshold equal to "0.05". Firstly, we create a folder named "Datasets" in the same directory that the Toolbox is located. Then, we simply create the tree hierarchy of files and folders as shown in Fig. 4 (left). If we run the Toolbox, we can load the option scenario by clicking on $Datasets \rightarrow myDataset \rightarrow 0.05 \rightarrow HS1.dat$, as shown in Fig. 4 (right). An option scenario can be also imported during runtime by following the same steps and by clicking $Datasets \rightarrow Update\ Datasets$.

**Fig. 4.** Automatic loading predefined option scenario.

# 6    Experimental Evaluation

We evaluated some of the implemented algorithms on real datasets, by using different parameters such as the number of sensitive itemsets to be hidden and the support count threshold. In this section, we also present the datasets used with their special characteristics, the selected parameters and the experimental results. All experiments were conducted on a PC running Windows 7 with an Intel Core i5, 3.20 GHz processor. For the linear programming techniques, CPLEX [2] was used for solving the formulated linear programs.

## 6.1  Datasets

All datasets used for evaluation are publicly available in the FIMI repository (http://fimi.ua.ac.be/data/). These datasets have different characteristics in terms of transactions and items, and the average transaction length. The characteristics of the datasets used are presented in Table 1. The mushroom dataset was prepared by Roberto Bayardo (University of California, Irvine) [6]. The retail dataset is a market basket dataset from an anonymous Belgian store [11]. The kosarak dataset was provided by Ferenc Bodon [9] and contains anonymized click-stream data of a Hungarian online news portal.

**Table 1.** Characteristics of the datasets.

| Dataset name | Number of transactions | Number of items | Avg. trans. length | $\sigma_{min}$ used |
|---|---|---|---|---|
| Mushroom | 8,124 | 119 | 23.00 | 1,625 |
| Retail | 88,162 | 16,470 | 10.30 | 22; 44; 66; 88 |
| Kosarak | 990,002 | 41,270 | 8.10 | 4,950 |

## 6.2   Evaluation Metrics and Framework

For the evaluation of the algorithms, we implemented and used several metrics along with the framework proposed by Bertino et al. [8]. The framework is based on several evaluation dimensions. We used the following metrics.

**Efficiency.** It is the ability of a PPDM algorithm to execute with good performance, in terms of all the resources consumed by the algorithm. Simply put, the efficiency of an algorithm quantifies how good is the relationship between its performance and the overall resources it uses. As in most cases, we assess efficiency in terms of time and space. In other words, efficiency is evaluated in terms of CPU time and the amount of memory that an algorithm requires.

**Scalability.** It is used to evaluate the behavior of the efficiency of a PPDM algorithm for a growing amount of input data, from which relevant information is mined while ensuring privacy. We conducted experiments with datasets of different size and density, so as to test the scalability of the algorithms implemented in the toolbox.

**Data Quality.** It refers to the quality of data after the hiding process. As mentioned earlier, attempting to hide sensitive information might have an impact on non-sensitive information as well. If data quality is too degraded, then the released database is useless for the purpose of knowledge extraction. According to Bertino et al. [8], the information loss can be measured in terms of the dissimilarity between the original dataset $D$ and the sanitized $D'$. The information loss is defined as the ratio between the sum of the absolute errors made in computing the frequencies of items in the sanitized database and the sum of all the frequencies of items in the original database.

We also use two additional measures: (a) the number of raw changes that occurred in data, and (b) the number of side effects. The raw data changes is the total number of items that have been removed in order to sanitize the database. The number of side effects (SE) introduced by the application of the sanitization process can be measured by $SE(\widetilde{F}, F') = |\widetilde{F}| - |F'| \geq 0$, where $|\widetilde{F}|$ is the number of itemsets in the revised set of frequent itemsets $\widetilde{F}$, whilst $|F'|$ is the number of itemsets in the set of frequent itemsets $F'$ mined from $D'$.

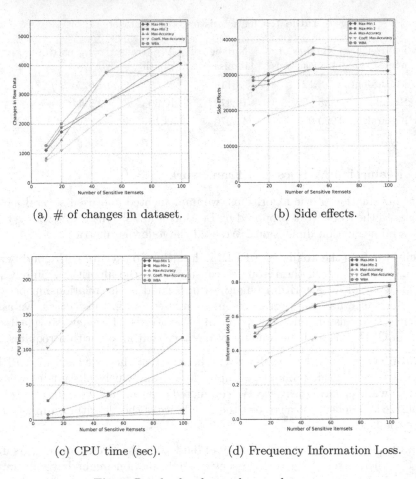

(a) # of changes in dataset.

(b) Side effects.

(c) CPU time (sec).

(d) Frequency Information Loss.

**Fig. 5.** Results for the mushroom dataset.

## 6.3   Experimental Results

Figure 5 presents the results we obtained for the mushroom dataset. Figure 5(a) displays how many changes (item removals) each algorithm made in the original database. Figure 5(b) displays the number of side effects that occurred as a result of the concealing process. Figure 5(c) presents the times needed by each algorithm. Lastly, Fig. 5(d) presents the frequency information loss. For the evaluation with this dataset, we used 4 different hiding scenarios; hiding 10, 20, 50 and 100 sensitive itemsets of different, random length. The support threshold used is $\sigma_{min} = 1625$. We selected randomly the sensitive itemsets.

The mushroom dataset is a small, yet dense dataset. Thus, the number of frequent itemsets increases dramatically as we decrease the support threshold. Linear programming techniques (Max-Accuracy and Coefficient-Based Max-Accuracy) achieve better results than their heuristic-based counterparts

(Max-Min 1, Max-Min 2 and WBA). Concerning the time complexity, we notice that the simpler the algorithm, the less time is needed to run, as expected.

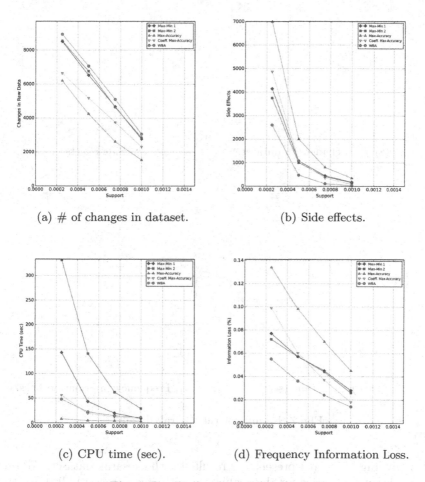

(a) # of changes in dataset.          (b) Side effects.

(c) CPU time (sec).          (d) Frequency Information Loss.

**Fig. 6.** Results for the retail dataset.

Figures 6(a)–(d) present the results for the retail dataset. We used a single hiding scenario of 100 sensitive itemsets. The set of sensitive itemsets consists of randomly selected itemsets. We performed experiments with this hiding scenario for different support thresholds, $\sigma_{min} = \{22, 44, 66, 88\}$.

The lower the mining threshold is, the larger the values of all metrics are. Notice that for this dataset, which is not as dense as the mushroom dataset, WBA achieves the best results, as far as side effects and information loss are concerned, with a fairly good time complexity. Although the results are printed in the text editor of the toolbox, the figures drawn in the canvas give a direct sense of which algorithm prevails.

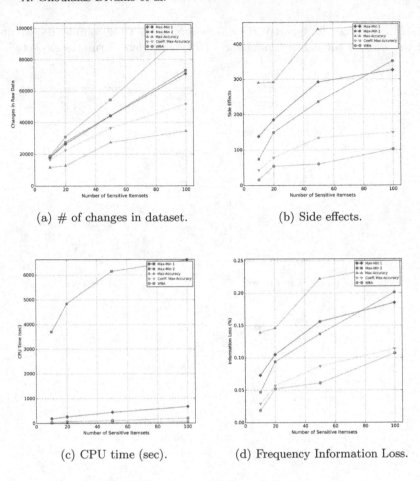

(a) # of changes in dataset.    (b) Side effects.

(c) CPU time (sec).    (d) Frequency Information Loss.

**Fig. 7.** Results for the kosarak dataset.

Finally, Figs. 7(a)–(d) present the results for the kosarak dataset. We used 4 different hiding scenarios for the evaluation with this dataset; hiding 10, 20, 50 and 100 sensitive itemsets of different, random length. The support threshold used is $\sigma_{min} = 4950$. The sensitive itemsets were picked randomly. Again WBA has the best results in terms of the number of side effects and the information loss. The time complexity is quite low for most of the algorithms and increases linearly with respect to the number of sensitive itemsets.

From the aforementioned experimental results, it is clear that the execution times of most of the techniques increase linearly as the number of sensitive itemsets increases, the size of the dataset increases, and the support threshold decreases. The density of the dataset has a great impact on the results. Linear programming techniques have good scalability. Then, the border-based techniques, such as Max-Min 1 and WBA, follow. Max-Min 2 appears to have a poor scalability compared to the rest of the heuristic algorithms.

# 7     Conclusions and Future Work

In this paper, we presented a FIH toolbox which can be used to apply a suite of hiding techniques on real world datasets. The toolbox comes with a built-in library containing several implementations of FIH algorithms and a suite of performance metrics. Currently the toolbox is in beta version and many improvements can be made, concerning both the GUI and the overall performance. Several features are currently under development, including a feature to recommend the appropriate FIH algorithm that, based on the characteristics of the input dataset, is expected to give the best results.

# References

1. https://github.com/kagklis/Frequent-Itemset-Hiding-Toolbox-x86
2. IBM ILOG CPLEX User's Manual v12.6
3. Agrawal, R., Srikant, R.: Privacy-preserving data mining. In: SIGMOD Conference, pp. 439–450 (2000)
4. Askari, M., Safavi-Naini, R., Barker, K.: An information theoretic privacy and utility measure for data sanitization mechanisms. In: Proceedings of the 2nd ACM Conference on Data and Application Security and Privacy (CODASPY 12), pp. 283–294 (2012)
5. Atallah, M., Bertino, E., Elmagarmid, A., Ibrahim, M., Verykios, V.: Disclosure limitation of sensitive rules. In: Proceedings of the 1999 Workshop on Knowledge and Data Engineering Exchange (KDEX 99), pp. 45–52 (1999)
6. Bayardo Jr., R.J.: Efficiently mining long patterns from databases. In: Proceedings of the 1998 ACM SIGMOD International Conference on Management of Data (SIGMOD 98), pp. 85–93 (1998)
7. Behnel, S., Bradshaw, R., Citro, C., Dalcin, L., Seljebotn, D.S., Smith, K.: Cython: the best of both worlds. Comput. Sci. Eng. **13**(2), 31–39 (2011)
8. Bertino, E., Fovino, I.N., Provenza, L.P.: A framework for evaluating privacy preserving data mining algorithms. Data Min. Knowl. Discov. **11**(2), 121–154 (2005)
9. Bodon, F.: A fast APRIORI implementation. In: Proceedings of the IEEE ICDM Workshop on Frequent Itemset Mining Implementations (FIMI 03), vol. 90, pp. 56–65 (2003)
10. Borgelt, C.: Frequent item set mining. Wiley Interdisc. Rev. Data Min. Knowl. Discov. **2**(6), 437–456 (2012)
11. Brijs, T., Swinnen, G., Vanhoof, K., Wets, G.: Using association rules for product assortment decisions: a case study. In: Proceedings of the 5th ACM SIGKDD International Conference on Knowledge Discovery and Data Mining (KDD 99), pp. 254–260 (1999)
12. Clifton, C., Marks, D.: Security and privacy implications of data mining. In: Proceedings of the 1996 ACM SIGMOD International Conference on Management of Data, pp. 15–19 (1996)
13. Clifton, C., Kantarcioğlu, M., Vaidya, J.: Defining privacy for data mining. In: National Science Foundation Workshop on Next Generation Data Mining (WNGDM), pp. 126–133 (2002)
14. Dasseni, E., Verykios, V.S., Elmagarmid, A.K., Bertino, E.: Hiding association rules by using confidence and support. In: Moskowitz, I.S. (ed.) IH 2001. LNCS, vol. 2137, pp. 369–383. Springer, Heidelberg (2001). https://doi.org/10.1007/3-540-45496-9_27

15. Evfimievski, A., Srikant, R., Agrawal, R., Gehrke, J.: Privacy preserving mining of association rules. Inf. Syst. **29**, 343–364 (2004)
16. Gkoulalas-Divanis, A., Verykios, V.S.: An integer programming approach for frequent itemset hiding. In: CIKM, pp. 748–757 (2006)
17. Gkoulalas-Divanis, A., Verykios, V.S.: Hiding sensitive knowledge without side effects. Knowl. Inf. Syst. **20**(3), 263–299 (2009)
18. Gkoulalas-Divanis, A., Verykios, V.S.: Exact knowledge hiding through database extension. IEEE Trans. Knowl. Data Eng. **21**(5), 699–713 (2009)
19. Johnsten, T., Raghavan, V.V.: A methodology for hiding Knowledge in databases. In: Proceedings of the IEEE International Conference on Privacy, Security and Data Mining (CRPIT 14), pp. 9–17 (2002)
20. Kagklis, V., Verykios, V.S., Tzimas, G., Tsakalidis, A.K.: An integer linear programming scheme to sanitize sensitive frequent itemsets. In: International Conference on Tools with Artificial Intelligence (ICTAI 14), pp. 771–775 (2014)
21. Leloglu, E., Ayav, T., Ergenc, B.: Coefficient-based exact approach for frequent itemset hiding. In: eKNOW2014: The 6th International Conference on Information, Process, and Knowledge Management, pp. 124–130 (2014)
22. Lindell, Y., Pinkas, B.: Privacy preserving data mining. In: Proceedings of the 20th Annual International Cryptology Conference on Advances in Cryptology (CRYPTO 00), pp. 36–54 (2000)
23. Menon, S., Sarkar, S., Mukherjee, S.: Maximizing accuracy of shared databases when concealing sensitive patterns. INFORMS **16**(3), 256–270 (2005)
24. Moustakides, G.V., Verykios, V.S.: A MaxMin approach for hiding frequent itemsets. Data Knowl. Eng. **65**(1), 75–89 (2008)
25. Oliveira, S.R., Zaïane, O.R.: Protecting sensitive knowledge by data sanitization. In: Proceedings of the 3rd IEEE International Conference on Data Mining (ICDM 2003), pp. 99–106 (2003)
26. Oliveira, S.R., Zaïane, O.R.: Algorithms for balancing privacy and knowledge discovery in association rule mining. In: IDEAS, pp. 54–65 (2003)
27. Oliveira, S.R., Zaïane, O.R.: Protecting sensitive knowledge by data sanitization. In: ICDM, pp. 613–616 (2003)
28. Pontikakis, E.D., Tsitsonis, A.A., Verykios, V.S.: An experimental study of distortion-based techniques for association rule hiding. In: Farkas, C., Samarati, P. (eds.) DBSec 2004. IIFIP, vol. 144, pp. 325–339. Springer, Boston (2004). https://doi.org/10.1007/1-4020-8128-6_22
29. Pontikakis, E.D., Verykios, V.S., Theodoridis, Y.: On the comparison of association rule hiding heuristics. In: Hellenic Database Management Symposium (2004)
30. Saygin, Y., Verykios, V.S., Clifton, C.: Using unknowns to prevent discovery of association rules. SIGMOD Rec. **30**(4), 45–54 (2001)
31. Stavropoulos, E.C., Verykios, V.S., kagklis, V.: A transversal hypergraph approach for the frequent itemset hiding problem. Knowl. Inf. Syst. **47**(3), 625–645 (2016)
32. Sun, X., Yu, P.S.: Hiding sensitive frequent itemsets by a border-based approach. JCSE **1**(1), 74–94 (2007)
33. Verykios, V.S., Elmagarmi, A.K., Bertino, E., Saygin, Y., Dasseni, E.: Association rule hiding. IEEE Trans. Knowl. Data Eng. **16**(4), 434–447 (2004)
34. Verykios, V.S., Pontikakis, E.D., Theodoridis, Y., Chang, L.: Efficient algorithms for distortion and blocking techniques in association rule hiding. Distrib. Parallel Databases **22**(1), 85–104 (2007)

# Author Index

Printed in the United States
By Bookmasters